CHEMISTRY *today*

CHEMISTRY *today*

The Portrait of a Science

ALFRED NEUBAUER

ARCO PUBLISHING, INC.
New York

Translated from the German by Sabine Scherer, Cambridge (Canada)
Revised by Jacek Klinowski, Cambridge (U.K.)

Published 1983 by Arco Publishing, Inc.
215 Park Avenue South, New York, N.Y. 10003
ISBN 0-668-05838-2
Copyright © 1982 by Edition Leipzig
Drawings by Gerhard Raschpichler and Rudolf Schilling
Designed by Egon Hunger
Jacket and cover design after a painting by Grant D. Venerable
Manufactured in the German Democratic Republic

Contents

Preface

The purpose of this book is to present to the general reader some of the most important developments and discoveries in chemistry, especially those made in the 20th century. I have tried to show the immense importance of chemistry in the development of human society by examining some of its effects and potential. Occasional digressions into the earlier history of chemistry have the purpose of illuminating the origin of certain approaches or developments. The scope of present-day chemistry reflected in its many disciplines, makes it very difficult to select the topics which can be treated adequately within the framework of such a book as this. I sincerely hope that the selection made will satisfy the interests of the reader.

Chemistry is a science which is intimately related to other fields of scientific endeavour. This is especially true of the relationship between chemistry and physics and between chemistry and biology. Since the end of the 19th century, the interrelatedness of chemistry and physics has decisively influenced the formulation of the theoretical fundamentals of chemistry. The nature of the relationship becomes clear as we consider the development of ideas about and definitions of such things as atoms, ions, molecules, chemical bonds, etc. We will discuss this topic in the introductory chapter as well as explaining a number of basic concepts necessary for the understanding of the chapters which follow.

The knowledge gained through chemical research, chemistry's modes of thought and its experimental methods (especially those of physical chemistry) have been and are indispensable for research into organic processes on the molecular level. This is clearly demonstrated by the contributions of biochemistry in the last three decades in clarifying the structure and function of proteins and nucleic acids.

The biochemical research was of truly revolutionary importance to biology. Since life is not possible without proteins or nucleic acids, I felt it appropriate to devote adequate space to these "highest ranking" chemical compounds.

The fact that the last 50 to 60 years have seen the development of polymer chemistry into a major field of research is reflected in the examination of a number of socially useful synthetic polymers besides our examination of the proteins and nucleic acids as the most important naturally occurring polymers.

Finally, I should like to thank all those who have helped me in obtaining the illustrations and through helpful comments on the manuscript. Thanks are also due to the staff of Edition Leipzig for their constructive assistance in the development and the editing of this book.

Alfred Neubauer

8

I.

Atoms, Ions, Molecules

Chemistry is the science which concerns itself with research into the properties and structures of substances and with the transformation of one kind of substance into another. Chemistry is interested in all types of matter, such as sand, iron, salt, sugar, water, diamond, air, cotton and nylon. Today, we know that all the things which surround us are compounded of about 90 basic forms of matter, the natural chemical elements. These elements are types of matter which cannot be reduced to simpler forms of matter through chemical reactions. If we assume purity, iron and diamond constitute such elements, diamond being a particular variation of the element carbon. Air is primarily a mixture of the elements oxygen and nitrogen, and it displays the characteristics of these two elements. Substances such as salt, sugar, water, cotton, and nylon, are made up of various elements and are called chemical compounds. The chemical compound water, for instance, the product of a chemical reaction between hydrogen and oxygen, displays entirely different properties than the original mixture of the two elements.

The Chemical Elements

When the justly renowned French chemist Antoine-Laurent Lavoisier (1743–1794) published his textbook, *Traité Elementaire de Chimie* in 1789, the table of elements in it consisted of only 21 elements. Today, we know of 107 elements for certain. They are assigned atomic numbers from 1 to 107.

The names of the elements are, as a rule, derived from Greek or Latin. The elements with the very high atomic numbers are largely named for famous scientists. Thus, element 99 was given the name "Einsteinium".

The introduction of abbreviation symbols for the elements by the Swedish chemist Jöns Jakob Berzelius (1779–1848) in 1814 became the basis for the modern international symbols for the elements. According to this, the elements are symbolized by their initials alone, or by their initial and one other letter from the name. Despite the fact that, at present, there are 107 known elements, Table 1 on pages 11 and 12 contains only 104 elements since, up till now (1978), there are no agreed upon names for the elements with atomic numbers higher than 104 (in the above sense of a name derived from that of a famous scientist). The difficulties and uncertainties of establishing elements with very high atomic numbers sometimes leads to conflicting claims of first discovery by several research teams. Such claims are reflected in the two names suggested for element 105, namely Nielsbohrium and Hahnium.

IUPAC, the International Union of Pure and Applied ·Chemistry, has proposed a systematic nomenclature for elements with higher atomic numbers than 100, thus providing a unified naming system. The name of the element is to be derived from the element's atomic number under the application of specific numerical word stems: 0 = nil, 1 = un, 2 = bi, 3 = tri, 4 = quad, 5 = pent, 6 = hex, 7 = sept, 8 = oct, 9 = enn. According to this system, the elements with the numbers 105, 106, 107 would become unnilpentium, unnilhexium, and unnilseptium respectively. Symbols of abbreviations for such element names would consist of three letters, e.g. Unh for unilhexium. This suggestion for a systematic nomenclature would not infringe upon the right of the discoverer of a new element to suggest another name once his discovery had been verified.

If we consider the actual occurrence of elements on our planet, we find that the quantitative distribution is highly uneven. The average quantities of the elements contained in the upper 10 miles of earth's crust are known. A ton of earth's crust, for instance, contains only 0.005 g gold, but 466,000 g oxygen and 277,200 g silicon.

The given relative atomic weight of the elements is based on the values known in 1975. The bracketed relative atomic weight values for certain radioactive elements[1] indicates the isotope[2] with the greatest half-life[3] or the atomic weight of best known isotopes.

Six elements, namely oxygen (O), silicon (Si), aluminium (Al), Iron (Fe), calcium (Ca), and sodium (Na), thus constitute 94% of the entire earth's crust. Not all of the elements mentioned in the Table of

1 Radioactive elements are chemical elements which decay due to the fact that they emit energy in the form of radiation.
2 Isotope: see page 26.
3 Half-life: the period of time in which half of a given amount of a radioactive element decays.

Table 1. The Chemical Elements

Element	Symbol	Atomic number	Relative atomic weight	Element	Symbol	Atomic number	Relative atomic weight
Actinium	Ac	89	227.0278	Iron	Fe	26	55.847
Aluminium	Al	13	26.98154	Krypton	Kr	36	83.80
Americium	Am	95	(243)	Kurtchatovium	Ku	104	257
Antimony	Sb	51	121.75	Lanthanum	La	57	138.9055
Argon	Ar	18	39.948	Lawrencium	Lr	103	(260)
Arsenic	As	33	74.9216	Lead	Pb	82	207.2
Astatine	At	85	(210)	Lithium	Li	3	6.941
Barium	Ba	56	137.33	Lutetium	Lu	71	174.97
Berkelium	Bk	97	(247)	Magnesium	Mg	12	24.305
Beryllium	Be	4	9.01218	Manganese	Mn	25	54.9380
Bismuth	Bi	83	208.9804	Mendelevium	Md	101	(258)
Boron	B	5	10.81	Mercury	Hg	80	200.59
Bromine	Br	35	79.904	Molybdenum	Mo	42	95.94
Cadmium	Cd	48	112.41	Neodymium	Nd	60	144.24
Caesium	Cs	55	132.9054	Neon	Ne	10	20.179
Calcium	Ca	20	40.08	Neptunium	Np	93	237.0482
Californium	Cf	98	(251)	Nickel	Ni	28	58.70
Carbon	C	6	12.011	Niobium	Nb	41	92.9064
Cerium	Ce	58	140.12	Nitrogen	N	7	14.0067
Chlorine	Cl	17	35.453	Nobelium	No	102	(259)
Chromium	Cr	24	51.996	Osmium	Os	76	190.2
Cobalt	Co	27	58.9332	Oxygen	O	8	15.9994
Copper	Cu	29	63.546	Palladium	Pd	46	106.4
Curium	Cm	96	(247)	Phosphorus	P	15	30.97376
Dysprosium	Dy	66	162.50	Platinum	Pt	78	195.09
Einsteinium	Es	99	(254)	Plutonium	Pu	94	(244)
Erbium	Er	68	167.26	Polonium	Po	84	(209)
Europium	Eu	63	151.96	Potassium	K	19	39.0983
Fermium	Fm	100	(257)	Praseodymium	Pr	59	140.9077
Fluorine	F	9	18.998403	Promethium	Pm	61	(145)
Francium	Fr	87	(223)	Protactinium	Pa	91	231.0359
Gadolinium	Gd	64	157.25	Radium	Ra	88	226.0254
Gallium	Ga	31	69.72	Radon	Rn	86	(222)
Germanium	Ge	32	72.59	Rhenium	Re	75	186.207
Gold	Au	79	196.9665	Rhodium	Rh	45	102.9055
Hafnium	Hf	72	178.49	Rubidium	Rb	37	85.4678
Helium	He	2	4.00260	Ruthenium	Ru	44	101.07
Holmium	Ho	67	164.9304	Samarium	Sm	62	150.4
Hydrogen	H	1	1.0079	Scandium	Sc	21	44.9559
Indium	In	49	114.82	Selenium	Se	34	78.96
Iodine	I	53	126.9045	Silicon	Si	14	28.0855
Iridium	Ir	77	192.22	Silver	Ag	47	107.868

Element	Symbol	Atomic number	Relative atomic weight		Element	Symbol	Atomic number	Relative atomic weight
Sodium	Na	11	22.98977		Tin	Sn	50	118.69
Strontium	Sr	38	87.62		Titanium	Ti	22	47.90
Sulphur	S	16	32.06		Tungsten	W	74	183.85
Tantalum	Ta	73	180.9479		Uranium	U	92	238.029
Technetium	Tc	43	(97)		Vanadium	V	23	50.9414
Tellurium	Te	52	127.60		Xenon	Xe	54	131.30
Terbium	Tb	65	158.9254		Ytterbium	Yb	70	173.04
Thallium	Tl	81	204.37		Yttrium	Y	39	88.9059
Thorium	Th	90	232.0381		Zinc	Zn	30	65.38
Thulium	Tm	69	168.9342		Zirconium	Zr	40	91.22

Table 2. The average amount of some of the elements contained in the earth's crust.

Element	g/t	Percentage		Element	g/t	Percentage
Oxygen	466,000.	46.60 %		Phosphorus	1,180.	.118 %
Silicon	277,200.	27.72 %		Sulphur	520.	.052 %
Aluminium	81,300.	8.13 %		Copper	45.	.0045 %
Iron	50,000.	5.00 %		Mercury	0.5	.00005 %
Calcium	36,300.	3.63 %		Gold	0.005	.0000005 %
Sodium	28,300.	2.83 %		Osmium	0.001	.0000001 %

Chemical Elements occur "naturally". Included in that list are also the artificially produced elements, the synthetic elements, such as Mendelevium or Kurtchatovium.

Dalton's Atomic Theory

Closely related to the concept of the "chemical element" is the concept of the "atom". The philosophic concept of an "atom" (*atomos* = Greek = the indivisible) has been in existence for two and a half thousand years. The concept and the related ideas were basic for the development of the materialistic nature philosophy of the ancient Greeks. The Greek philosophers Leucippus (ca. 500—440 B.C.) and Democritus (ca. 460–370 B.C.) developed the hypothesis that the material world, i.e. all substances, ultimately consists of final indivisible particles, namely atoms. The joining or separation of atoms of different kinds would account for the differences between individual things. However, these theories did not explain the exact nature of these constituent atoms.

Nowadays, we know a number of smallest particles of matter for whom the concept of further divisions has become experimentally irrelevant, and which are, in this sense, indivisible. These particles, also known as elementary particles, could be called atoms in the original sense of the word, had chemistry not already appropriated the term at the beginning of the 19th century. At that time, the English scientist John Dalton (1766–1844) developed the first scientific atomic theory on the basis of quantitative experiments with chemical compounds. Dalton's work became the decisive theoretical foundation for the further development of chemistry.

John Dalton, British scientist, 1766–1844.

The most important theses of Dalton's Atomic Theory were:

- All chemical elements are made up of small indivisible particles, known as atoms. Atoms have weight.
- The atoms of a particular element are identical but differ from the atoms of all other chemical elements.
- A chemical compound comes into existence when atoms of two or more chemical elements bond together, forming a stable unit.

Dalton stated that the atoms which constitute the elements have weight, but in his lifetime there was no scientific method for determining the absolute atomic weight. In the 19th century, however, it became possible to determine the so-called *relative atomic weights* of the elements. An analysis of the weight ratios of elements in various compounds made it possible to establish that the hydrogen atom is the lightest of all atoms. Thus, the hydrogen atom was given the atomic weight 1 by definition, and the weights of all the other elements were determined relative to this unit. Thus, chlorine had an atomic weight of about 35.5 which simply means that an atom of chlorine weighs 35.5 times as much as an atom of hydrogen. Relative atomic weights are, therefore, merely symbols of a relationship. More recently, relative atomic weights have been designated as *relative atomic mass*. According to this modern definition, the relative atomic mass of an element is its mass expressed relative to 1/12th of the mass of the atom of the carbon isotope ^{12}C. An understanding of this definition requires previous acquaintance with the concept of an isotope, a concept which we will explain later (see p. 26). Table 1, "The Chemical Elements", gives exact figures corresponding to the relative atomic weights resp. mass in column 4.

Mendeleyev's System of the Elements

The number of known elements kept on increasing during the 19th century. When chemists began to examine the characteristics of particular elements and compare them with each other, they realized that some elements showed considerable similarities in their chemical and physical characteristics. They began to sort similar elements into groups, organizing the elements within the groups according to their relative atomic weights. The various attempts to organize the elements into a system were crowned with success in 1869, in the then published paper "A System of the Elements, arranged according to their Atomic Weights and Functions" by the Russian chemist Dmitri Ivanovich Mendeleyev (1834–1907). Independent of Mendeleyev's work, the German chemist, Lothar Meyer (1830–1895), had arrived at similar conclusions. Whereas Mendeleyev's system first appeared in German chemistry journals in the spring of 1869, Meyer published his system in December of the same year.

Periodic System of the elements
(the long period representation)

The values given above the element symbols correspond to the relative atomic weights. The numbers below the element symbols are the atomic numbers of the elements. The bracketed values for the relative atomic weight of certain radioactive elements refer to the isotope with the longest half-life.

1st period	1,0079 H 1														
2nd period	6,941 Li 3	9,01218 Be 4													
3rd period	22,9898 Na 11	24,305 Mg 12													
4th period	39,0983 K 19	40,08 Ca 20	44,956 Sc 21												
5th period	85,47 Rb 37	87,62 Sr 38	88,905 Y 39												
6th period	132,905 Cs 55	137,34 Ba 56	138,91 La 57	140,12 Ce 58	140,907 Pr 59	144,24 Nd 60	[145] Pm 61	150,35 Sm 62	151,96 Eu 63	157,25 Gd 64	158,924 Tb 65	162,50 Dy 66	164,930 Ho 67	167,26 Er 68	168,934 Tm 69
7th period	[223] Fr 87	226,03 Ra 88	227,03 Ac 89	232,038 Th 90	231,03 Pa 91	238,03 U 92	237,05 Np 93	[244] Pu 94	[243] Am 95	[247] Cm 96	[247] Bk 97	[251] Cf 98	[254] Es 99	[257] Fm 100	[258] Md 101

The manner in which Mendeleyev presented his system of the elements in 1869 ordered similar elements in horizontal groupings, like the element group fluorine (F), chlorine (Cl), Bromine (Br), and iodine (I).

Fig. 2. shows a modern version of Mendeleyev's System of the Elements. Mendeleyev himself laid special emphasis on the periodicity of the elements, i.e. the recurrence of characteristics at discrete intervals on a scale where the elements were organized according to their increasing atomic weight. The elements were grouped into "periods". In Mendeleyev's System of 1869 these periods are represented by six vertical columns. In 1871, however, Mendeleyev introduced the horizontal representation of the periodic table, the system with the so-called "short periods". At present, the most usual representation of the table of the elements is in the form of long periods; the first period containing two elements, the second and third periods eight elements each, the fourth and fifth periods eighteen elements each, the sixth period contains thirty-two elements. The seventh period is incomplete.

Dmitri Ivanovich Mendeleyev, Russian chemist, 1834–1907.

																	4,0026 He 2
											10,81 B 5	12,011 C 6	14,0067 N 7	15,9994 O 8	18,9984 F 9	20,179 Ne 10
											26,9815 Al 13	28,086 Si 14	30,9738 P 15	32,06 S 16	35,453 Cl 17	39,948 Ar 18
	47,90 Ti 22	50,942 V 23	51,996 Cr 24	54,9381 Mn 25	55,847 Fe 26	58,9332 Co 27	58,71 Ni 28	63,546 Cu 29	65,38 Zn 30	69,72 Ga 31	72,59 Ge 32	74,9216 As 33	78,96 Se 34	79,904 Br 35	83,80 Kr 36
	91,22 Zr 40	92,906 Nb 41	95,94 Mo 42	[97] Tc 43	101,07 Ru 44	102,905 Rh 45	106,4 Pd 46	107,870 Ag 47	112,40 Cd 48	114,82 In 49	118,69 Sn 50	121,75 Sb 51	127,60 Te 52	126,9044 I 53	131,30 Xe 54
173,04 Yb 70	174,97 Lu 71	178,49 Hf 72	180,948 Ta 73	183,85 W 74	186,2 Re 75	190,2 Os 76	192,2 Ir 77	195,09 Pt 78	196,967 Au 79	200,59 Hg 80	204,37 Tl 81	207,19 Pb 82	208,980 Bi 83	[209] Po 84	[210] At 85	[222] Rn 86
[259] No 102	[260] Lr 103	[257] Ku 104	105	106	107											

The System of the Elements, presented in 1869 and 1871, showed a number of gaps. Not surprisingly, since a considerable number of elements had not as yet been discovered. Mendeleyev predicted the characteristics of a number of yet undiscovered elements on the basis of the law of periodicity. In 1871, for instance, he predicted the existence of an element—he called it eka-silicum—with the approximate atomic weight of 72 and described several of its physical and chemical properties. In 1886, his prediction was experimentally confirmed with the discovery of germanium. The discoverer of this element, the German chemist Clemens Winkler (1838–1904), wrote: "There is no doubt that the newly discovered element is the eka-silicum described by Mendeleyev fifteen years ago. A more impressive confirmation of the validity of the laws of periodicity is hard to imagine..." Sensational successes of this sort helped to achieve general recognition for the System of the Elements, which had originally been accepted with considerable reservations.

The organization of elements according to a rising scale of atomic weights involves difficulties in assigning certain elements to their respective groups. The element potassium (K), for instance, with its atomic weight of 39.0983 should really change places with the element argon (Ar) since argon has the higher atomic weight of 39.948. Potassium would then fall under the element neon in the periodic table. However, the alkaline metal potassium and the noble gas neon have nothing in common, so the principle of organization according to rising atomic weight was by-passed, and potassium was placed after argon. The deeper reason for this difficulty can be found in the fact that the ordering of the elements according to the characteristic "atomic weight" does not represent the best organizational principle. It was not until the elements were ordered according to their so-called nuclear charge (atomic number) that the contradiction disappeared. This did not become possible until scientists had learned more about the structure of the atom.

Particle Models of the Atom

In the 19th century the atom was conceived of as a spherical, solid particle, evenly filled with matter[1] and perfectly elastic. Its behaviour was considered to be in accordance with the laws of classical mechanics. This mechanical model was replaced at the end of the 19th and the beginning of the 20th centuries by a new model which

16

1 Matter: the word is here used in the sense of a weighable substance.

tried to include an explanation for the "electrical properties" of the atom. According to this model, atoms were conceived as spheres which were evenly filled with mass and which carried a positive electric charge in which were embedded unmoving smaller spheres with a negative electric charge (electrons).

The theory that atoms are compact, matter filled spheres finally had to be abandoned on the basis of evidence from the experiments of Rutherford[1] in 1910–11. By irradiating metal foils with alpha rays (nuclei of helium atoms) Rutherford showed that atoms are largely "empty" and that their mass is almost entirely concentrated in the positively charged nuclei. This led to the Rutherford model, which is based on the following postulates:

— Atoms contain a positively charged nucleus whose radius is of the order of 10^{-5} nm. Almost the entire mass of the atom is concentrated in the nucleus.
— The atomic nucleus is surrounded by an extremely loose shell consisting of electrons whose total negative charge is equal to the positive charge of the nucleus.
— The electrons revolve about the atomic nucleus in orbits of arbitrary radius and angle. The stability of the atom is achieved through the equilibrium between the electrostatic attraction of the electrons to the nucleus and the centrifugal force of the electron motion.

Rutherford's model represented a considerable advance in the understanding of the structure of the atom. All the same, it left several problems unsolved. Most importantly, it contradicted the laws of electrodynamics. An orbiting electron, i.e. a moving electric charge,

17

Ernest Rutherford (Lord Rutherford), 1871–1937.

The Rutherford model of the atom. The positively charged nucleus N is circled by electrons (e⁻) in orbits of arbitrary angle and radius.

The Bohr model of the atom. The electrons move strictly on "predetermined" orbits. If an electron makes a quantum jump from an orbit of higher energy to an orbit of lower energy, then the atom emits energy.

1 Rutherford, Ernest, 1871–1937, British physicist. Nobel Prize for Chemistry in 1908.

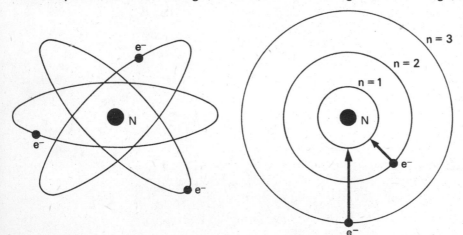

Niels Bohr, 1885–1962.

1 Bohr, Niels, 1885–1962,
Danish physicist. Nobel Prize
for Physics in 1922.
2 Planck, Max 1858–1947,
German physicist. Nobel
Prize for Physics in 1918.
3 Newton, Isaac, 1643–1727,
British physicist, mathematician and astronomer.

should create an electromagnetic alternating field which, in turn, should give off energy. An electron which continually emits energy should lose velocity, should be more and more strongly attracted by the positively charged nucleus, and should finally fall into the nucleus on a spiral path. However, none of this happens. Was the model wrong? Or did the laws of electrodynamics not apply in the micro-world of the atoms?

As early as 1913, Bohr[1] attempted to solve the apparent contradiction by using the Quantum Theory which had been developed by Planck[2]. In 1900, Planck had proposed the thesis which was to revolutionize physics. He postulated that radiation consists of indivisible, very small portions of energy which he called quanta. Energy of the quanta is dependent on the frequency of the particular radiation source. However, there is such a thing as a minimum amount of energy, its smallest possible portion, which cannot be reduced further. This theory contradicted the then prevalent concept that radiation energy was like a stream and could be absorbed or emitted in any amounts, no matter how small. Despite the fact that Newton[3] had already developed a particle theory of light, the belief that radiation consisted of electromagnetic waves had prevailed. Not until the arrival of the Quantum Theory and its experimental confirmation at the beginning of the 20th century did the particle theory of light re-emerge. At the same time, the wave theory remained in force, so that light, in our present-day thinking, is a physical object, which, when looked at with the help of certain instruments, provides observational data consistent with wave properties. In certain other experiments, the observational data can only be interpreted by the particle theory of light. Thus, we speak of the "dual nature" of light. However, light does not consist of waves under some circumstances and of particles under others. It must be strongly emphasized that light remains one and the same physical entity. It is we who are unable to see it as a unified phenomenon. Not every physical manifestation provides us with a unified conceptual model. That which we have just said of light is valid for the entire range of radiation phenomena.

The concept of energy quanta was used by Bohr in order to extend the Rutherford model of the atom. Bohr postulated that the electrons of the atom do not circle the atomic nucleus on arbitrary paths but in certain predetermined orbits only. Furthermore, Bohr assumed that the electrons revolve without emitting any sort of energy on the predetermined orbits which correspond to their particular energy states. An atom only emits energy when an electron shifts from an orbit of higher energy (more distant from the nucleus) to an orbit of lower energy (closer to the nucleus). These transitions from one permissible energy state to another are known as *quantum jumps*.

With the help of Bohr's atomic model, it was now possible to explain the existence of line spectra for monoatomic metal vapours and noble gases. According to this theory, the spectra correspond to the different energy states of the electrons. Conversely it was possible to draw conclusions as to the energy states of the electrons in the atoms by analyzing the atomic spectra. Bohr analyzed the line spectra of the hydrogen atom in the light of the model. His results were confirmed experimentally a few years later. According to Bohr, the hydrogen atom consists of a positively charged nucleus which is orbited by a single electron on one of certain concentric paths. The orbit with the smallest radius (r = 0.053 nm) where the electron can usually be found, also represents the lowest state of energy, the "ground state", of the electron. If more than one electron circles the nucleus of an atom in a single orbit, we speak of an electron shell.

A further quantum mechanical refinement of Bohr's model, largely worked out by the physicists Sommerfeld[1] and Pauli[2], resulted in the thesis that the location and the energy level of every electron in the shell of an atom can be described by four quantum numbers: the principal quantum number n, the angular momentum quantum number l, the magnetic quantum number m, and the spin quantum number s. Theoretically, the principal quantum number can assume any positive integral value, from 1 to infinity. All electrons with the same principal quantum number are located in the same principal shell. The angular momentum quantum numbers characterize the energy differences between electrons in the same principal shell. These numbers can have values from 0 to n − 1; i.e. if n = 2, then the angular momentum quantum number of the electrons will be 0 and 1. By tradition, electrons with the angular momentum quantum number 0, 1, 2, 3, are referred to as s-, p-, d-, f-electrons.

Two rules apply:

1. In one atom no two electrons can ever have identical quantum numbers.
2. Every principal shell can contain a maximum of $2n^2$ electrons. For example, the shell with the principal quantum number n = 1 can have a maximum of $2 \times 1^2 = 2$ electrons. If n = 2, the maximum number of electrons in the shell would be $2 \times 2^2 = 8$.

Given these two principles, the systematic combining of the four

1 Sommerfeld, Arnold, 1868–1951, German theoretical physicist.
2 Pauli, Wolfgang, 1900–1958, Austrian physicist. Nobel Prize for Physics in 1945.

quantum numbers will yield information as to all possible energy states of the electrons in the atomic shells. It is thus possible to characterize the electron configurations of atoms theoretically. This, in turn, yields a theoretical basis for the periodic system of the elements. However, before we return to the periodic system, we will take a further look at the atomic model and up-date it further. Up to now, we have described the particle model of the electron. 20th century physics has, however, shown that electrons possess wave properties as well.

The Wave-Mechanical Model of the Atom

In 1924 the French physicist de Broglie[1] postulated the following: "The electrons cannot be regarded simply as particles of electricity; it must be considered to have wave properties as well."

The wave properties of electrons were confirmed experimentally in 1927. Using a beam of fast-moving free electrons it was possible to produce diagrams analogous to those produced with X-rays, a form of radiation whose wave properties had already been established. According to de Broglie, the electron has the form of a circular standing wave. Bohr's postulated, but unexplained, thesis that only certain orbits are permitted to the electrons, could now be justified. Electron orbits can only be such that their circumference corresponds to an integral multiple of the wavelength of the electron. Otherwise, overlapping would cause interferences or even obliteration of wave patterns. The existence of these "standing waves" is the condition of the quantification of the electrons.

Bohr's other thesis, that electrons on these certain orbits rotate without emitting energy, could now also be explained: a self-contained circular wave cannot emit energy. The mathematical fundamentals for the electron-wave theory were primarily worked out by Schrödinger[2] in his development of a wave mechanics. In 1926 he defined the electron in the hydrogen atom as a standing spatial wave which encloses the nucleus, and proceeded to describe this wave mathematically in the form of a wave equation (the Schrödinger Equation).

A further fundamental discovery at the end of the 1920s led to the abandonment of Bohr's concept of the electron, considered as a particle which circles the atomic nucleus in specific orbits and at specific velocities. This discovery was Heisenberg's[3] Uncertainty Principle, formulated in 1925, which constitutes the basic law of behaviour for

1 De Broglie, Louis Victor, born 1892. Nobel Prize for Physics in 1929.
2 Schrödinger, Erwin, 1887–1961, Austrian physicist. Nobel Prize for Physics in 1933.
3 Heisenberg, Werner, 1901–1976, German physicist. Nobel Prize for Physics in 1932.

objects of the atomic order of magnitude. According to this position and momentum in the world of sub-atomic particles can never be determined simultaneously with an arbitrary degree of accuracy. Since momentum is the product of the mass and velocity of a moving body, the uncertainty principle can also be stated in other terms: The position and momentum of elementary particles cannot be measured simultaneously with arbitrary degree of accuracy. The more precisely the position of an elementary particle is determined, the less accurate will be a simultaneous measurement of its velocity. This means that it is not possible, in principle, to make the precise statements concerning the orbits and the speeds of electrons as could be derived from Bohr's model of the atom. Even Schrödinger's solutions on wave equations cannot provide information about the exact position of an electron at a given time. It is, however, possible to use quantum mechanics in order to arrive at statements about the probability with which one is likely to encounter an electron at a particular place and time. The greater the amplitude of a wave at a particular point, the greater the probability that the electron will be found there. The space in which the probability of the electron's presence is above a certain definite value is called its *orbital*. To be more precise, orbitals are the space in the atom with 90 per cent probability of electron presence. In order to make the concept of the wave-mechanical nature of an electron a little clearer, the picture of a *charge cloud* whose shape corresponds to the limits of the orbital can be

21

Louis Victor de Broglie, born 1892.

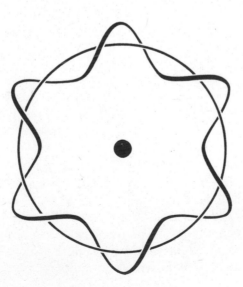

The wave-mechanical model of a hydrogen atom with the electron as a standing wave, in accordance with de Broglie's hypothesis.

Werner Heisenberg,
1901–1976.

used. Since wave mechanics describes electrons as pulsating electromagnetic fields distributed around the atomic nucleus in the form of spatial standing waves, it is also reasonable to think of electrons as charge clouds arranged around the nucleus. Depending upon the energy states of the individual electrons in the atoms, defined by the quantum numbers, charge clouds—or orbitals—of varying sizes and shapes are formed. In order to describe the various orbitals, the principal quantum numbers and the angular momentum quantum numbers are used. For the angular momentum quantum number, it is usual not to use numbers but instead to use the "old" letter convention s, p, d, f; for instance, there are orbitals such as 1s or 2p. If one includes the particle concept in one's picture of the electron, then it is possible to visualize the following illustration—which is a metaphor and not a "real picture"—of the charge cloud analogy. The cloud can be regarded as the sum of countless photographic snaps of the fast moving electron. The cloud's density is not uniform. There are areas where the electron can be found frequently or with a high probability (characterized in the picture by a higher density of dots).

There are also areas where the electron can only be found infrequently (denoted by a low density of dots) or where it cannot be found at all (no dots). The charge clouds or orbitals do not have clearly defined outer limits. In principle, they are infinitely large, which contradicts our actual experience that atoms are of a finite size. Since we conceive of orbitals as spaces in which there is 90 per cent probability of an electron presence, the matter becomes practicable, and we arrive at a finite, if uncertain, limit to the orbitals. In this manner, the calculated 1s orbital diameter is approximately 0.08 nm.

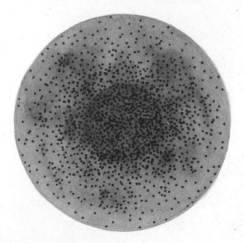

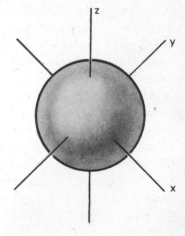

Cross-section of the charge
cloud of an electron.

Model representation of the
spherical 1s orbital.

This value is a close match with the indirectly derived diameter of the hydrogen atom, whose single electron occupies a 1s orbital in its ground state.

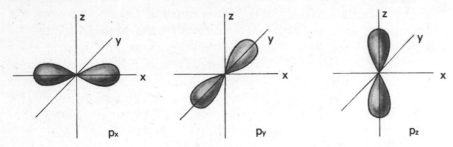

Model representation of 2p orbitals.

What specific spatial shapes can be assigned to the orbitals? The orbital with the lowest energy state, the so-called 1s orbital, is a sphere with the atomic nucleus at the centre. The next higher energy state orbital, 2s, is also spherical; it is merely larger than 1s orbital. There follow the three 2p orbitals of equal energy, whose spatial configuration is similar to that of an hourglass. The three 2p orbitals differ from each other by their spatial distribution around the nucleus. If we place the nucleus at the centre of a right-angled (cartesian) set of coordinates, one of the 2p orbitals is located on the x axis, one on the y axis, and the third on the z axis. These 2p orbitals are also known as $2p_x$, $2p_y$, and $2p_z$ orbitals.

Particularly pertinent in relation to the atomic orbitals is the principle (formulated by Pauli, see p. 19) according to which an atomic orbital cannot contain more than two electrons. These two electrons differ minimally in their energy level. The difference is due to a variance in their rotation velocity, the *spin* of the two electrons. We visualize the difference between the two electrons, by assigning to them opposite directions of rotation. Accordingly, one electron spins counter-clockwise, the other clockwise. When constructing the electron shells of different atoms, it must be remembered that the orbitals which correspond to higher energy states are only filled once the lower energy state orbitals are completely occupied. The total distribution of an atom's electrons between different orbitals is called its electron configuration. The higher the principal quantum number of an electron, the higher its energy. Electrons which have the same principal quantum number are located on the same principal quantum shell. The laws governing the quantum numbers dictate that the shell with the principal quantum number 1 can have only two electrons. This shell is identical with the 1s orbital. If there is only one electron in the 1s orbital, we have the electron shell of the simplest atom, that of hydrogen. If the 1s orbital is occupied by two electrons, we have the helium atom. After this, the principal quantum shell with the principal number $n = 2$ begins to be filled. As stated before, shells with the principal quantum number 2 may contain up to 8 electrons since the second principal quantum shell governs four orbitals. These are the 2s orbital and the three 2p orbitals. If these orbitals as well as the 1s orbital are fully occupied, we have the electron configuration of the neon atom[1]. It is here that the relationship to the periodic system (see pp. 14/15) becomes clear: every period in the periodic system corresponds to a principal quantum shell.

The principal quantum shell incorporates a particular energy state of an electron or several electrons; it also governs the lower energy levels such as s-, p-, d-, or f-, (the angular momentum symbols) which may involve one or more orbitals. Fig. 2 shows the relationship between the periods of the system of elements and the principal quantum shells of electrons with their subject energy levels. The latter and their orbitals are arranged in accordance with the increase in the value of their energy levels. Orbital 1s is the orbital with the lowest energy level; the three 7p orbitals represent the highest energy level. The occupation of orbitals of equal energy by electrons, for example the three p-orbitals, proceeds in the following manner: First, each 2p orbital is filled with one electron; then, and only then, does the filling of the second place for an electron in the orbital take place. Fig. 2 explains this law. The order of the occupation of the orbitals of the second principal quantum shell, and thus the order of elements in the second period of the periodic table is Li, Be, B, C, N, O, F, Ne.

The Nucleus of the Atom

In our concern with the negatively charged electron shell, let us not forget the positively charged atomic nucleus. Its positive charge is carried by sub-atomic particles which we call *protons*. All the atomic nuclei, with a single exception, also contain electrically neutral particles, the so-called *neutrons*. The existence of these uncharged nuclear particles was first suspected by Rutherford in 1920, but experimental confirmation was not obtained until 1932 through the research of the British physicist Chadwick[2]. Only the nucleus of the hydrogen atom

1 Neon has an $1s^2 2s^2 2p^6$ electron configuration. Upper case numbers represent the number of electrons per energy level.
2 Chadwick, James, 1891–1974, British nuclear physicist. Nobel Prize for Physics in 1935.

contains no neutrons. The nuclear building blocks, the neutrons and the protons, are collectively known as *nucleons*. Table 3 lists some of the data concerning all three atomic particles, the proton, neutron and electron.

Table 3

Name	Symbol	Mass	Charge
Proton	p	1.6726×10^{-24} g	positive
Neutron	n	1.6749×10^{-24} g	neutral
Electron	e^-	0.91096×10^{-27} g	negative

The mass of the neutron is only very slightly greater than that of the proton. In comparison with the electron, however, these particles both have almost 2,000 times as much mass. The mass of the atom can thus fairly be said to be almost exclusively determined by its protons and neutrons.

The simplest atom is the hydrogen atom which consists of one proton (nucleus) and one electron (shell). Sodium atoms have a more complicated structure, having a nucleus made up of 11 protons and 12 neutrons and a shell of 11 electrons. The electrical neutrality of the atom is always maintained, which means that the number of electrons in the shell always equals that of the protons in the nucleus.

An ordering of the various types of atoms (= elements) is achieved by sorting them in terms of the increasing number of protons in the nucleus. The resultant number is known as the atomic number (see Table 1, The Chemical Elements). A chemical element can thus be defined as a substance which is constituted of atoms with the same atomic number. The problem of the ranking of potassium and argon in the periodic table of elements, mentioned on p. 16, can now be solved without contradiction. Argon, with its 18 protons, has a lower atomic number than potassium with its 19 protons and thus naturally takes precedence in the Table of Elements.

The sum of protons and neutrons in an atom is called the mass number of the element. This mass number corresponds largely to the relative atomic weights, a phenomenon which has a very simple explanation. Originally, chemists assigned the relative atomic weight 1 to the lightest of the atoms, namely the hydrogen atom, whose mass consists almost entirely of a single proton.

The fact that the relative atomic weight or mass of the atoms frequently deviates somewhat from whole numbers was partly explained by experiments carried out in 1912 by the British physicist J. J. Thomson[1]. With the help of physical methods, Thomson establĩshed that the ordinary element neon, a noble gas, consists to approximately 90 % of neon atoms with the relative atomic weight of 20, and to approximately 10 % of neon atoms with the relative atomic weight of 22. In reality, there are three kinds of neon with the relative atomic weights of 20, 21 and 22 with 90.92, 0.257 and 8.82 % of neon. It is the mixture of all three that gives a relative atomic weight of 20.179.

The same "mixture" fact holds true for most of the elements. There are "pure" elements which really consist of only one type and weight of atom, such as fluorine, sodium, aluminium, iodine, and gold. The reason for the deviating mass numbers in the mixed elements is a variation in the number of neutrons in the atomic nucleus. The element chlorine, wherever it is found on this earth, has a mass number of 35 for 74.4 % of its atoms and the mass number 37 for the remaining 24.6 %.

Both types of atoms always have 17 protons and 17 electrons each, but the number of neutrons in the nucleus of the first type of chlorine atom is 18 and in the second type, it is 20. Since these two different atoms retain the same atomic number and differ only minimally if at all in their chemical behaviour, they also occupy the same place in the periodic system. This is the origin of the term that refers to them in their difference–*isotope*–which comes from the Greek words *isos* meaning "the same" and *topos,* meaning "place". In chemical nomenclature, isotopes are characterized by writing the mass number of the particular atom in upper case, the atomic number in lower case, in front of the symbol for the atom. For instance, $^{12}_{6}C$ describes a carbon atom with a mass number of 12. The absolute mass of a carbon atom ^{12}C equals $19.926372 \cdot 10^{-24}$ g. In 1961, $^1/_{12}$ of the absolute mass of the carbon isotope ^{12}C was agreed upon as the standard for the relative atomic mass of the elements (see p. 12).

The Chemical Bonds

Even though two chemists (Otto Hahn and Fritz Strassmann[2]) discovered in 1938 how to split the nucleus of the uranium atom, the study of atomic nuclei is properly a part of the field of physics.

1 Thomson, J. J., 1856–1940, British physicist. Nobel Prize for Physics in 1906.
2 Strassmann, Fritz, 1902–1980, German chemist.

Since the formation of chemical compounds requires the bonding together of atoms and since this process always involves a change in the electron shells of the atoms involved, the chemists are primarily interested in all the activities and processes taking place in the electron shell of atoms.

Ionic Bonds

If electrically neutral atoms lose electrons or gain additional ones in their electron shells, then positively or negatively charged atoms result. These charged atoms are called *ions*. If the atom is positively charged, we speak of *cations*, if negatively charged, of *anions*. Ions with opposite charges exert a mutual attraction upon each other and, given a certain proximity, they may bond together, forming an ionic bond. In principle, atoms can absorb additional electrons as long as they have orbitals with an incomplete complement of electrons. For example, the chlorine atom has 17 electrons in its electron shell which are distributed in the following manner: $1s^2 2s^2 2p^6 3s^2 3p^5$. This means that one of the 3p orbitals has room for an extra electron since a full complement for the 3p orbitals would mean $3p^6$. The absorption of an extra electron provides the chlorine atom with 18 electrons. These 18 negative charges are (now no longer) balanced by only 17 positive charges (protons) in the nucleus, so that we now have a simple negatively charged ion, called the chloride ion.

27

Otto Hahn, German chemist, 1879–1968, Nobel Prize for Chemistry in 1944.

$$Cl \quad + \quad e^- \quad \rightarrow \quad Cl^-$$
chlorine atom + electron → chloride ion
(negative)

Such chemical reactions, in which the reacting particles (atoms, ions, groups of atoms) absorb electrons are called reduction reactions.

If we ask what neutral atom in the periodic table of the elements naturally possesses 18 electrons, we find the noble gas argon, which is located right next to chlorine in the table. Noble gases, as a rule, have fully occupied quantum shells. Such full electronic shells are very stable and therefore react minimally with other elements. The extremely reactive and unstable chlorine atom is clearly stabilized by its acquisition of an extra electron, a process by which it achieves the very stable electron configuration of the noble gas argon.

There are other ways for atoms to achieve the electron configuration of a noble gas. If we take the sodium atom with its electron con-

Crystal lattice of sodium chloride.

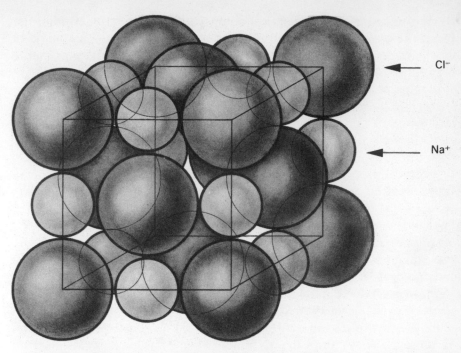

← Cl⁻

← Na⁺

figuration $1s^2 2s^2 2p^6 3s^1$, we find, immediately preceding it in the periodic table, the noble gas neon with its electron configuration of $1s^2 2s^2 2p^6$. To change its electron configuration to that of the neon atom, the sodium atom merely needs to shed the $3s^1$ electron, a matter that is not difficult in energy terms. The giving-off of this electron causes the proton number in the nucleus to be one-too-many, which means that the resultant sodium ion is positively charged.

$$Na \rightarrow Na^+ + e^-$$

sodium atom → sodium ion + electron
(positive)

Chemical reactions which involve the loss of an electron are known as oxidation reactions.

The coupling of an oxidation and a reduction reaction is known as a redox reaction. An example of such a redox reaction is a part of the chemical reaction between sodium atoms and chloride atoms which results in sodium chloride, better known as table salt.

$$Na \rightarrow Na^+ + e^- \quad \text{oxidation}$$
$$e^- + Cl \rightarrow Cl^- \quad \text{reduction}$$

$$Na + Cl \rightarrow Na^+ Cl^- \quad \text{redox reaction}$$

In the process, the electrons which are freed through the oxidation of the sodium are absorbed by the chlorine which becomes chloride ion. The resultant ions, the Na^+ and the Cl^- are electrostatically attracted to one another. They stick together, forming the chemical compound sodium chloride. The ionic bond is especially prevalent among the salts. In the crystals of table salt, the Na^+ and the Cl^- are arranged in an orderly manner in set places. They form a crystal lattice made up of oppositely charged ions.

Metallic Bonds

Most of the chemical elements are metals. Metals, and the alloys formed with them, in their solid state are built of crystal lattices. The difference between these lattices and the ionic lattices is that the entire lattice is occupied by positive metal ions. Between these cations there can be found a number of more or less freely moving electrons given off by the outer quantum shell of the metal atoms. These free electrons neutralize the electrostatic repulsion forces between the positive metal ions and effect a bond between them. It is these electrons that have given rise to the metaphor of an "electron gas". The model is particularly apt since it also explains certain properties of metals, such as electrical and thermal conductivity. All the same, it is an extremely rough approximation of the true relations in the crystal lattices of the metals. The "electron gas" of the metals has properties that differ considerably from those of a normal gas.

The application of the orbital concept to metallic bonds led to a derivation of metal orbitals, which are described as consisting of occupied and free atomic orbitals in the metal atoms and which extend across many atoms in the crystal lattice. These metal orbitals are very close to each other in their energy level so that a clear distinction into discrete energy levels is no longer possible. The energy levels merge into a band of continuous energy distribution in which the electrons move freely (band model).

Covalent Bonds

We shall not pursue theoretical problems involved in metallic bonding further. Instead, we shall take a closer look at the form of bonding which has received most attention from the chemists and physicists:

covalent or *homopolar bonding*. An example of a covalent bond is the bonding of two hydrogen atoms into an hydrogen molecule. Molecules are defined as independent or relatively independent particles which consist of at least two atoms, either identical or different. The atoms in the molecule are held together by covalent bonds. Sodium chloride in its solid state is thus not constituted of molecules since, in the first place, there is no covalent bonding and, in the second, there are no independent NaCl particles which are arranged in a crystal lattice but Na^+ or Cl^- ions. An aqueous solution of sodium chloride does not contain NaCl particles but Na^+ and Cl^- ions, moving freely in the water and separated from one another by the water molecules.

Using our knowledge about the atomic orbitals we will now try to illustrate the formation of a covalent bond. The bonding of two hydrogen atoms into a hydrogen molecule will serve as an example.

$$H + H \longrightarrow H_2$$

hydrogen atoms   hydrogen molecule

As we know, the hydrogen atoms possess a 1s orbital occupied by only one electron. When a bond is formed between two hydrogen atoms, there is a partial overlapping of these two orbitals. The more the orbitals overlap, the stronger is the covalent bond. The two electron clouds form a new single cloud, a molecular orbital, which is occupied by two electrons. Each electron can move within the area of the entire bonding orbital but the electron density is naturally most pronounced in the area of the overlap. The prerequisite for the formation of such a molecular bonding orbital is an opposite spin on the part of the electrons involved. Only if this condition is met can the formation of an electron pair which will occupy a molecular orbital take place. Like atomic orbitals, molecular orbitals can only contain two electrons per orbital.

In the process of covalent bonding between two hydrogen atoms into one hydrogen molecule, the two positive nuclei approach each

The formation of a molecular orbital. Separate 1s orbitals of two hydrogen atoms.

Overlapping 1s orbitals of two hydrogen atoms.

H

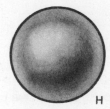

H

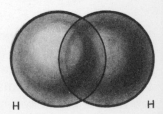

H   H

other until their electrostatic repulsion (particles with the same charge repel each other) prevents a further approach. The part of the electron cloud which lies between the two nuclei attracts the nuclei to itself and is thus responsible for their approaching one another in the first place. Thus the electrons function as the "glue" which counteracts, at least partly, the nuclei's forces of repulsion, at least to some extent. The result of this interaction between the atomic nuclei and the electrons is the formation of the molecule. The arrangement of the electrons and the atomic nuclei in the new molecule shows a lower state of energy than the two original atoms. The energy content of this system of two hydrogen atoms is at its lowest if the two atomic nuclei are at a distance of 0.074 nm from each other, the length of the covalent bond in the hydrogen molecule.

Long before chemists knew anything about covalent or ionic bonds[1], to say nothing of orbitals, they used the dash in their formulae to symbolize the bond between atoms. Since a novel representation of molecular orbitals as the bond between two atoms would create unnecessary complications in chemical nomenclature, chemists to-day continue to use the simple dash, which they have re-defined to mean a bonding or common electron pair. Dots are also used to symbolize electrons. H_2 is the summary formula for the hydrogen molecule; H:H or H−H are the two representations of the structural formula for this molecule. The summary formula merely gives information as to the type and number of atoms in a molecule. The structural formula also show the order in which the various atoms are bonded to each other and the type of bond involved. Additionally, structural formulae can reflect the spatial organization of the atoms in the molecule. If the bond between two atoms consists of a single electron pair, then we speak of a single bond. The water molecule (H_2O), for instance, contains two single bonds between the oxygen atom and the two hydrogen atoms:

The two horizontal lines at the top and the bottom of the oxygen atom symbolize "free electron pairs" which are not involved in the bonding.

The structural formula shows that the atoms in the water molecule are not arranged in a line but at an angle. The bond angle is 104.5°.

1 The existence of these two types of bonds was postulated in 1916. The German physicist Walther Kossel (1888–1956) developed the theory of ionic bonds and the American physical chemist Gilbert Newton Lewis (1875–1946) the theory of covalent bonds.

The theory of the orbitals serves to explain the nonlinear structure of the water molecule. The electron configuration of the oxygen atom is $1s^2 2s^2 2p^4$. The occupation of the three 2p orbitals takes the following form: ⊙ ⊙ ⊙ i.e. one of the 2p orbitals is fully occupied by an electron pair; the other two orbitals are occupied by unpaired electrons (the circles symbolize the orbitals, the dots the electrons in the orbitals). These two unpaired electrons each enter into a covalent bond with the 1s electron in each of the two hydrogen atoms. For simplicity, we will assume that the electrons involved are p_x and p_z electrons. Since the 2p orbitals are vertical to one another, the overlap with the 1s orbitals results in the following spatial arrangement:

32 *Orbital model of a molecule of water H_2O.*

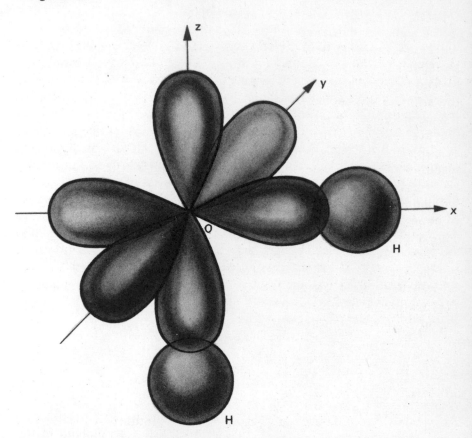

According to this figure, the H–O–H angle should be 90°. The fact that it is a little larger in reality is attributed to the repulsion forces between the two hydrogen atoms and the bonding electron pairs.

The nitrogen atom (electron configuration $1s^2 2s^2 2p^3$) has three un-paired 2p electrons which bond with three hydrogen atoms in order to form the ammonia molecule NH_3. According to the orbital theory, the $H-N-H$ angle should be 90°. Experimentally, however, a value of 107.5° has been determined. The theoretical value of 90° for the $H-S-H$ angle in the hydrogen-sulphide molecule (H_2S) is in close agreement with the experimentally determined 92°.

S—H   N—H
| / \
H   H   H

Despite the angle deviation from the theoretical values of the angle in the molecules of water and ammonia, the predictive power of the orbital theory concerning the spatial organization of these simple molecules is entirely satisfactory. Whereas the atoms of the molecules of hydrogen, water, and hydrogen-sulphide lie on a single plane, the ammonia molecule has a three-dimensional, pyramidal structure. If we consider the electron configuration of the atoms in the molecules that we have been discussing, we can see that, counting the shared electrons twice, the atoms have managed to achieve the very stable electron configurations of noble gases. Hydrogen atoms in these mole-cules possess the electron configuration of helium; the other atoms, oxygen, sulphur, and nitrogen, all achieve the electron configuration of neon.

In order to penetrate even deeper into the mysteries of covalent bonding, we will examine a molecule which possesses four simple bonds. The molecule which we will consider is the combustible part of natural gas, the methane molecule. With a summation formula of CH_4, we are given a very simple organic compound[1] whose structural formula is usually written in the following manner:

H
|
H—C—H
|
H

This symbolic representation does not reflect the spatial conformation accurately; it does, however, show the four single $C-H$ bonds. In order to form these four equivalent bonds we need four unpaired

1 Organic chemistry, in principle, is the chemistry of all carbon compounds. In-organic chemistry deals with all other elements and com-pounds apart from those containing carbon. However, some carbon compounds are considered inorganic. Examples are carbon dioxide CO_2 and sodium carbonate Na_2CO_3. The historical devel-opment of chemistry gave rise to this rather irrational division.

electrons in the carbon atom. The electron configuration of the carbon atom is $1s^2 2s^2 2p^2$ which indicates only two unpaired atoms:

2p²   ⊙ ⊙ ○← empty 2p orbital
2s² ⊙
1s² ⊙

Given this information, one could, for instance, predict the existence of a compound such as CH_2—and this compound does in fact exist, but it is a very reactive and unstable molecule which constantly seems to strive towards two further simple bonds.

How do we get four unpaired atoms in the carbon atom? The process has been theoretically explained as follows. To begin with, one of the 2s electrons is "promoted" to the empty 2p orbital. After this promotion, the electron configuration of the carbon atom is as follows:

2p³   ⊙ ⊙ ⊙
2s¹ ⊙
1s² ⊙

The energy for the promotion of the 2s electron to this higher state of energy is drawn from the energy that is released by the later formation of the four single bonds. However, up to now, we only have four unpaired electrons with very different energy levels—and the experimentally established equivalence of the four atomic bonds in the methane molecule requires energetic equivalence of all four electrons. The four unpaired equivalent orbitals are achieved by a mixing (hybridization) of the 2s orbital with the three 2p orbitals. This results in four mixed or *hybridized orbitals* with an equivalent energy level. Since one s orbital has combined with three p orbitals the newly formed hybrid orbitals in this case are sp^3 orbitals. The described process brings us to the following electron configuration for the carbon atom:

2p³ orbitals   ⊙ ⊙ ⊙ ⊙
1s² orbital   ⊙

Such sp^3 orbitals are considered to have the following spatial arrangement:

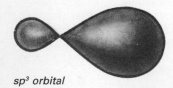

sp³ orbital

The four sp³ orbitals of the carbon atom are arranged spatially as to be at the greatest possible distance from each other. The spatial distribution which corresponds to this requirement is one directing the four orbitals into the corners of a tetrahedron[1]. In the formation of the methane molecule, the sp³ orbitals overlap with their thicker ends on the 1s orbital of the hydrogen atoms, so that an orbital model of these four bonds looks as follows:

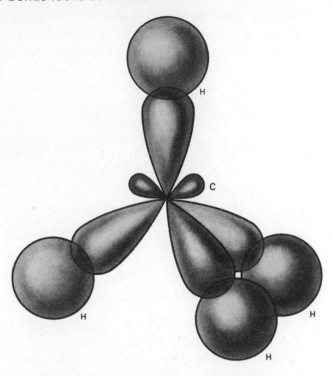

Orbital model of the four C—H bonds in methane.

The H—C—H bonding angles in methane measure 109.5° and the C—H bonds all have the same length of 0.11 nm.

In comparison with the ionic bonds, which exert their force equally in all directions (an ion surrounds itself with as many oppositely charged ions as possible), the atomic or covalent bonds have directional properties which determine the molecular structure.

The conceptual model of the carbon tetrahedron for certain organic compounds was postulated intuitively more than a hundred years ago, in 1874, by a twenty-two year old Dutch chemist, van't Hoff. In early September of that year he published a paper (in Dutch) entitled *Suggestions for the Expansion of the Presently Used Structural Formulae in Chemistry into a Spatial Dimension, plus Related*

1 Tetrahedron is a regular polyhedron with four faces.

Notes as to the Relationship between the Optical Rotation Capacity and the Chemical Constitution of Organic Compounds. With this paper, van't Hoff became one of the founders of that branch of chemistry which concerns itself with the spatial distribution of atoms in molecules, the so-called *stereochemistry*[1]. After his work had been translated into French (1875) and German (1876), van't Hoff's ideas spread rapidly. Their propagation was catalyzed by the ruthless denigration of these ideas by one H. Kolbe[2] who damned the theory of the young Dutch chemist in the following terms:

Jakobus Henricus van't Hoff, 1852–1911, received the first Nobel Prize for Chemistry in 1901.

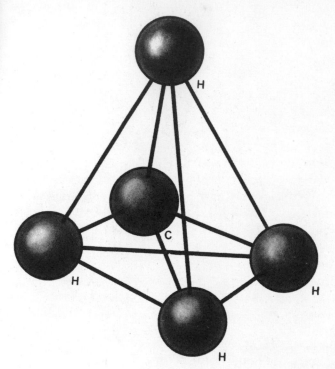

Tetrahedral model of a methane molecule according to van't Hoff.

1 In the last quarter of the 19th century, the French chemist Le Bel (1847–1930) made important contributions to the development of stereo-chemistry. In November 1874 he published, independent of van't Hoff's research, a paper with the title *Sur les relations qui existent entre les for-*

''... a Dr. J. H. van't Hoff, an employee of the veterinary college in Utrecht, finds, it seems to us, exact chemical research little to his taste. Instead, he has found it more comfortable to bestride Pegasus (no doubt on loan from the veterinary college) and to publish in his *La chimie dans l'Espace* his visions of the organization of the atoms in space, which clearly appeared to him on his flight up the Parnassus of chemistry... Such a treatment of serious scientific questions, not far removed from faith in witchcraft and spiritualism, is actually considered admissible by Wislicenus.''[3]

However, van't Hoff and his theory rose, like the proverbial phoenix, from the scorching phrases of this attack. He was victorious because his stereochemical ideas were correct and proved extraordinarily fruitful in the further development of organic chemistry.

The development of the tetrahedron model is closely associated with the concept of the asymmetric carbon atom. This refers to a carbon atom to which four different atoms or atom groups are bonded in the tetrahedral configuration. For such compounds, van't Hoff stated: "...that, in the case where four affinities of a carbon atom are joined to four different univalent groups, two and only two different tetrahedra can be formed, one of which is the mirror image of the other. However, it is important to remember that the two can never be made congruent. They constitute two isomeric structural formulae in space..."[1]

Let us clarify this basic thesis of van't Hoff by means of our hands. Our hands have this same relation of image and mirror image to each other. No matter how hard we try, we can never superimpose them. They are not congruent in the proper sense; though alike, they are inverted, left to right. The compounds with asymmetric carbon atoms are molecules, which, like our hands, exist in the form of image and mirror image and cannot be made congruent despite all their similarity. The following illustration shows model representations of the image and mirror image molecules of such a compound. Model I and Model II cannot be made congruent, no matter how hard one tries. The reader can test this to his own satisfaction with homemade models. This phenomenon is a case of *isomerism*. Isomeric compounds have the same summary formula but they differ in the sequence of their atoms or in the spatial arrangement of the atoms. In the above case, the compound represents a special kind of spatial or *stereo-isomerism*. Such stereo-isomeric molecules are also known as enantiomers. Thus we have "left-handed" and "right-handed" molecules. The property has even been called "handedness". However, the term in most common use, derived from the Greek, is chirality. The left-handed molecules are designated by an L (Latin: *laevus* = left) and the right-handed ones with a D (Latin: *dexter* = right). If a mixture or compound consists of both D and L molecules in equal amounts, then we have a racemic form which is denoted by a DL before the name of the molecule.

The D- and L-forms of a chemical compound have the same colour and the same melting point. As a matter of fact, nearly all their physical properties are identical. There is, however, one physical prop-

mules atomiques des corps organiques et le pouvoir rotatoire de leurs dissolutions (Concerning the relations which exist between the chemical formulae of organic compounds and their rotational capacity in solutions). Le Bel dealt with stereo-chemical problems much like those of van't Hoff but did not make the tetrahedron model the centre of his considerations.
2 Kolbe, Hermann, 1818–1884, taught in Leipzig, an influential organic chemist.
3 *La chimie dans l'espace* (Chemistry in Space), the French title of van't Hoff's paper.
Pegasus: the winged steed of Greek mythology.
Parnassus: a mountain range in Greece; in antiquity it was reputed to be the home of the muses and this later turned it into a symbol of poetry.
Wislicenus, Johannes, 1835–1902, a German organic chemist who propagated van't Hoff's ideas.

1 affinities: bonds;
univalent: one-valued.

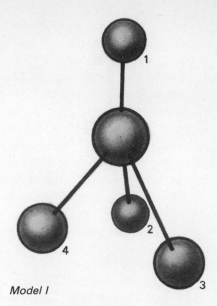

Model I

38

erty in which the mirror image isomers differ: they rotate polarized light in different directions. Normal white light consists of a mixture of rays of different wavelengths which vibrate in different planes. One can produce light consisting of only one wave-length and vibrating only on a single plane. This is known as polarized light. When polarized light passes through a solution of a chiral compound, the plane of polarization is shifted by a certain amount—and it will be shifted the same amount in the opposite direction if the isomer is exchanged for one of the opposite "handedness". This property of molecules with chirality, the ability to produce a shift in the plane of polarized light, is known as its optical activity. Of course, racemic (DL) mixtures do not rotate polarized light since the opposing effects of the two isomers cancel each other. A classical example of a chiral compound is lactic acid.

Assigning elements to either the D- or the L-group of isomers is not done in accordance with the way in which the isomer shifts polarized light (right or left). Instead, compounds are assigned to one of the groups by way of a comparison with certain standard compounds whose spatial structure is clearly understood. A molecule with one asymmetric carbon atom can exist in two mirror image isomeric forms. A molecule with two asymmetric carbon atoms can take four (2^2) mirror image isomeric forms. With ten asymmetric carbon atoms, there is a theoretical possibility of 1,024 mirror image isomeric forms (2^{10}).

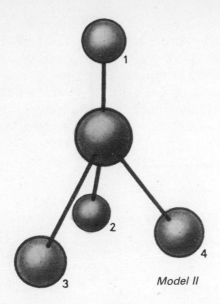

Model II

Research on chiral compounds is by no means only a pretty mental game for the inhabitants of a scientific ivory tower. Instead, it plays an important practical role in the synthesis of various medicinal drugs. Fine details in the spatial arrangement of chemical compounds can be of decisive importance for their biological effectiveness. The stereochemistry of the α-amino acids and the proteins clearly demonstrates this (see chapter IV, The Molecular Foundations of Life).

Multiple Bonds

Apart from the single bonds which we have been discussing up to this point, there are also double and triple bonds between atoms, which are formed by two or three common electron pairs respectively. It is one of the basic explanations for the truly immense number of organic compounds that carbon atoms can enter into all three kinds of bonds. For example, the compound ethane consists of a carbon-carbon single bond; ethene or ethylene contains a carbon-carbon double bond, and ethyne or acetylene a carbon-carbon triple bond. van't Hoff conceived of a C—C single bond as two tetrahedra with a shared corner. If the compound contains a double bond, then the two tetrahedra have two common corners, i.e. a common edge. A triple bond involves tetrahedra which share a face.

In modern terminology, the C–C single bond is described as the result of the overlapping of two sp³ orbitals of the two carbon atoms. It is characteristic of C–C single bonds that rotations can take place about this bond. This means that in ethane, for instance, the CH_3-groups rotate about this bond. We do not want to go into a detailed description of the formation and appearance of the molecular orbitals of ethene and ethyne at this point. Suffice it to say that the ethene

Summary and structural formulae of ethane, ethene and ethyne.

40

Compound	Summary formula	Structural formula
ethane	C_2H_6	H₂C—CH₂ (with H's around each C)
ethene	C_2H_4	H₂C=CH₂ (with H's around each C)
ethyne	C_2H_2	H—C≡C—H

bonds are formed between the carbon atoms through the overlapping of two sp² orbitals for the first bond; the second bond is due to the overlapping of two 2p orbitals. The second bond is a so-called π-bond (pi). In ethyne, one of the bonds is due to the overlapping of sp orbitals whereas the other two are π-bonds. All bonds which are formed through the overlapping of sp orbitals or sp³ or sp² orbitals are called σ (sigma) bonds. Since the overlapping of the p orbitals is not as strong as that of the sp³, sp² and sp orbitals, the π-bonds are weaker than the σ-bonds. This means that the π-bonds are also easier to break; i.e. they are more reactive. Finally, let us describe the spatial arrangement of ethene and ethyne. All the atoms of ethene lie on a

single plane. The $C \overset{\cdots H}{=} C$ angles are 121.5°; the $\overset{H}{\underset{H}{>}}C$ angles

are 117°. Ethyne, on the other hand, is a linear molecule. The H–C≡C angles are all 180°.

ethane

ethene

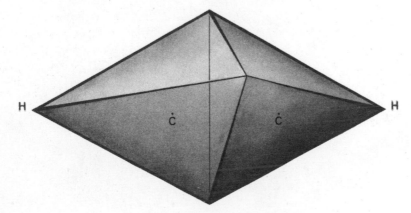

ethyne

Tetrahedral models of ethane, ethene and ethyne molecules according to van't Hoff.

The Aromatic Compounds

π-electrons also occur in the so-called aromatic compounds. A representative of this class of compounds is benzene, C_6H_6. As far back as 1865, Kekule[1] developed two equally valid structural formulae for benzene based on the tetravalent carbon atom. His structural formulae are still in use today.

Benzene formulae according to Kekulé.

42

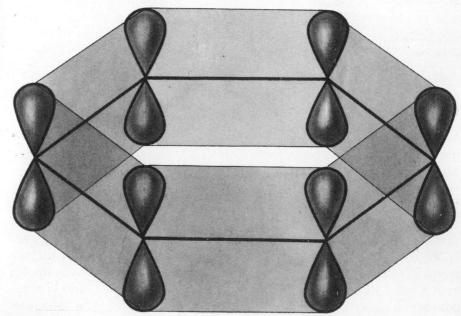

abbreviated form:

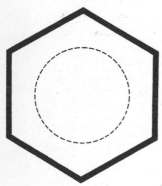

The modern formula for benzene.

The π-electron sextet in benzene.
The grey ground between the six 2p orbitals indicates their overlapping.

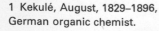

1 Kekulé, August, 1829–1896, German organic chemist.

Since it was not possible to describe the chemical behaviour of benzene unambiguously with these two formulae, the two structures have generally been interpreted as the limiting structures of this compound.

The true bond is therefore considered a mixture of the two limiting structures. Both formulae suggest an alternating double and single bond in ring form. However, in reality there is no alternation of bonds. The overlapping of the total of six π-electrons (in the three double bonds) creates a uniform circular π-electron system which is responsible for this new type of bond. The six π-electrons in benzene are symbolized by a dotted circle within the carbon framework.

Van der Waals Forces

Ionic, covalent and metallic bonds rely on the existence of strong interacting forces. Between atoms and molecules in both solid and liquid state there are also weak interacting forces known as van der Waals forces[1]. These are responsible, for instance, for the fact that noble gas atoms, hydrogen molecules, or non-polar methane molecules coalesce in sufficiently low temperatures, forming liquids or crystalline solids. The bigger the electron system of an atom or molecule, the larger the van der Waals forces between them. There is a rather complicated explanation concerning the existence and the nature of these forces. We will not dwell upon them in this volume.

The Hydrogen Bond

Among the weak interactive forces or bonds, we also count the hydrogen bond which can occur both *between* molecules (intermolecular) and *within* molecules (intramolecular).

A symmetrical distribution of bonding electrons is only achieved in molecules which consist of one sort of atom. The hydrogen molecule is an example of such a molecule. In a water molecule, the symmetry is already disturbed. The oxygen atom has exerted a greater attraction on the bonding electron pair than have the hydrogen atoms. We say that the oxygen atom has a greater electronegativity than the hydrogen atom. Other atoms, such as C, N, Cl, and F, also possess a greater electronegativity than hydrogen. The unequal

[1] Named after van der Waals, Johannes Diderek, 1837–1923, Dutch physicist, Nobel Prize for Physics in 1910.

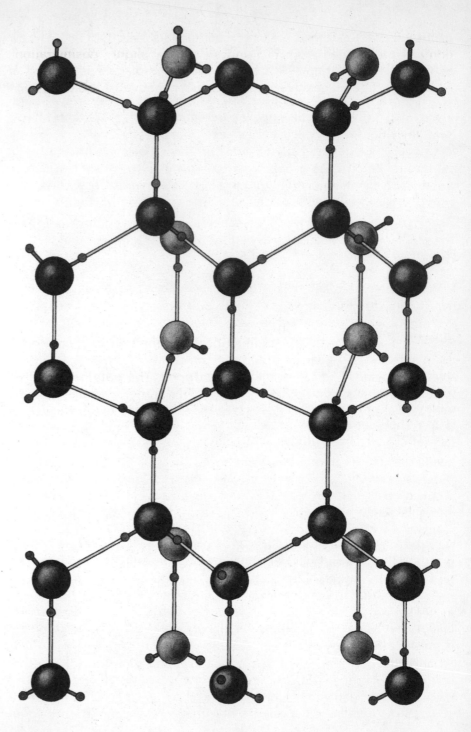

Hydrogen bonds determine the tetrahedral spatial structure of crystalline water (ice). The black and grey circles denote the oxygen atoms of the water molecules on the different levels of the ice crystal. The light, smaller circles denote the hydrogen atoms.

distribution of the bonding electron pair results in a slight "negativization" of the oxygen atom (δ^-) and an equally slight "positivization" of the hydrogen atoms (δ^+). In this way, a molecule is created of which one side is positive and the other negative: a molecule with two poles, a *dipolar molecule*.

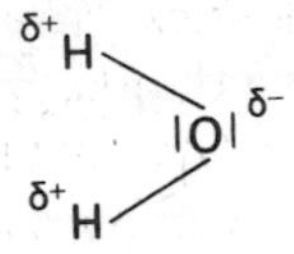

$$
\begin{array}{c}
{}^{\delta^+}H \\
 \diagdown \\
 |O|\,{}^{\delta^-} \\
 \diagup \\
{}^{\delta^+}H
\end{array}
$$

According to the principle that opposite poles attract one another, the dipolar water molecules develop a particular kind of bond:

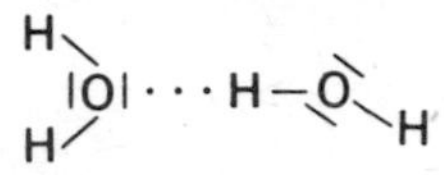

$$
\begin{array}{c}
H \\
\diagdown \\
|O| \cdots H - O \\
\diagup \diagdown \\
H H
\end{array}
$$

Since a hydrogen atom is the link between the two oxygen atoms, or in general, between any two electronegative atoms X —H ... Y, this type of bond is called a *hydrogen bond*. The dots between the water molecules symbolize this type of bonding. Hydrogen bonds can be found both in the liquid and the solid state (ice) of water. In liquid water, the hydrogen bonding effects the aggregation of 50 to 100 molecules; in ice, all the molecules are organized into tetrahedric spatial structures due to these hydrogen bonds. When ice melts, this three-dimensional structure is largely destroyed and many molecules move closer together, which in turn explains why liquid H_2O (water) has a higher density than solid H_2O (ice). It so happens that water reaches its maximum density at 4°C and thus ice floats in water. Hydrogen bonds also play a decisive role in the formation of particular structures in the proteins and the nucleic acids, the key substances in the chemistry of life (see Chapter IV, The Molecular Foundations of Life).

Models of Chemical Structures

In order to illustrate the structure of compounds, both structural formulae and models are used by chemists. Fig. 8, for example, shows van't Hoff's tetrahedron models. The models reflect, more or less accurately, the geometry of chemical compounds (bond angles, bond lengths, sizes of the atoms). There are "ball and stick" models in which the balls represent the atoms and the sticks the bonds. Skeleton models merely represent the bonding framework of the molecules. In frequent use are models in which the interacting spheres of the various atoms are represented by spatially extensive bubbles (see illustrations). We will use pictures of these various models in this book in order to illustrate particular structures.

1 Experiments in A. L. Lavoisier's laboratory involving the respiratory system of the human being, drawn by Madame Lavoisier. Lavoisier's explanation of the basic processes involved in combustion and in respiration were of inestimable importance for the further development of chemistry in the 18th century. Both processes are oxidation reactions. Moreover, Lavoisier made significant contributions to the definition of chemical elements and the systematization of the then known elements and compounds.

2   The relationship between the periods of the System of Elements and the principal quantum shells of the electrons, shown with their sublevels (in the 4th period the Si should be Sc, in the 5th period I instead of J, in the 7th period Lr instead of Lw).

3   In December 1938 Otto Hahn and Fritz Strassmann in Berlin discovered how to split uranium atoms with this apparatus. Cleavage of the atomic nucleus was achieved through irradiation with slow neutrons.

48

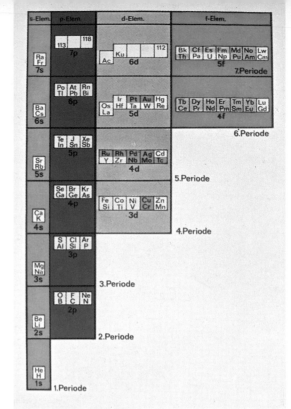

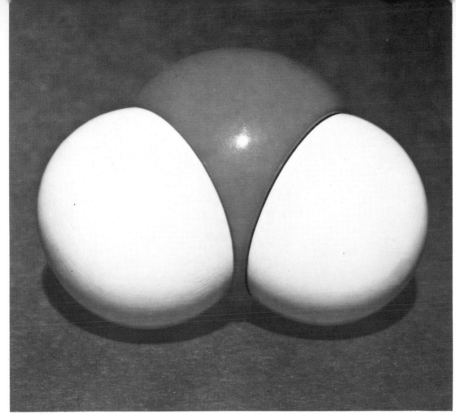

4   Space-filling model of
a water molecule.
Blue bubble = oxygen atom.

5   Space-filling model of an
ammonia molecule.
White bubbles = hydrogen
atoms.

49

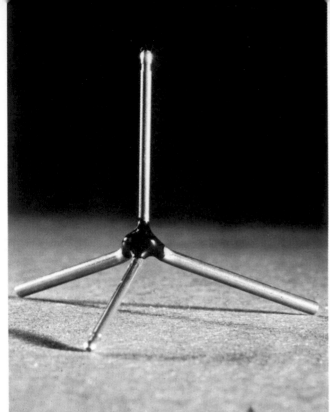

6   Skeleton model of methane.
The four metal rods represent
the four covalent bonds in
the methane molecule.

7   Space-filling model of
methane.
Black bubble = carbon atom;
white bubbles = hydrogen
atoms.

8   Tetrahedral molecular
models which were built by
J. H. van't Hoff (1852–1911)
and used by him in his work.

9/10 Skeleton models of the ecliptic (9) and the staggered (10) conformation of ethane. The two methyl groups ($-CH_3$) of the ethane can rotate about the C−C single bond. The different molecule shapes which result from this are known as molecular conformation. If the two methyl groups are in the same plane, then we speak of an ecliptic conformation. This conformation is not advantageous in energy terms since the arrangement forces the hydrogen atoms to approach one another so closely that they obstruct one another. If one of the methyl groups in the ecliptic conformation is turned in relation to the other by 60°, then the hydrogen atoms are staggered. The staggered conformation is the energetically preferred one.

11/12 *Space-filling models of the ecliptic (11) and the staggered (12) conformation of ethane.*

13 *Space-filling model of ethene.*

14 *Space-filling model of ethyne.*

15 *Space-filling model of buta-1,3-diene.*

16 *Space-filling model of benzene.*

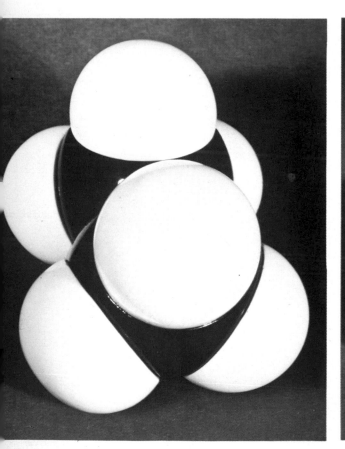

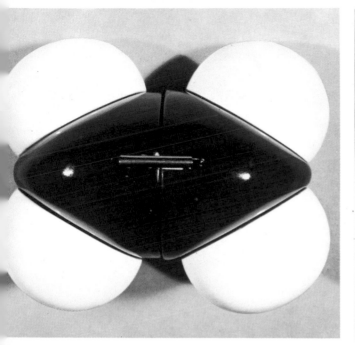

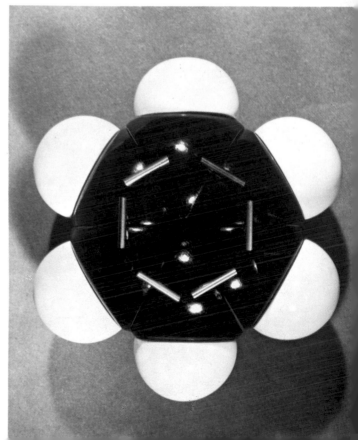

17 Electron-microscopic photograph of the fracture surface of a sodium chloride crystal (enlarged 15,000 times). The black specks in the photograph are gold decorations.

54

II

Recent Developments in Chemical Synthesis

In general, the concept of synthesis in chemistry refers to the artificial production of chemical compounds. It is also possible to synthesize elements, but since the synthesis of an element always involves nuclear reactions it belongs in the realm of nuclear physics. Among the artificially produced elements are technetium (atomic number 43), promethium (atomic number 61) and all the transuranium elements (atomic numbers 93–107 inclusive). The term synthesis is also used to refer to the production of one allotropic form of an element from another. An example is the production of diamond from graphite, which we will describe in detail in this chapter.

The synthesis of chemical compounds aims to produce naturally occurring substances on the one hand and at the creation of substances which are not found in nature on the other. Most of the known chemical compounds, of which there are more than 7 million at present, are the creations of chemists. Even now, the journals of chemistry regularly carry reports, over 250,000 per year, of new compounds synthesized in chemical laboratories. Although only a small percentage of the known compounds are produced in any considerable amount, these few serve to produce highly differentiated consumer products which contribute greatly to the satisfaction of numerous individual and social needs of modern man.

The basic types of reaction which lead to chemical compounds are the following:

- The reaction of an element with another element:
  $$C + O_2 \rightarrow CO_2$$
- the reaction of an element with a compound:
  $$CH_2 = CH_2 + Br_2 \rightarrow CH_2Br - CH_2Br$$

— the reaction of a compound with itself:
$$nCH_2 = CH_2 \xrightarrow{\text{polymerization}} \ldots -CH_2 - CH_2 - CH_2 - CH_2 - \ldots$$
— the reaction of a compound with another compound:
$$CH_3COOH + C_2H_5OH \rightarrow CH_3COOC_2H_5 + H_2O.$$

In the chemistry of natural substances the so-called total or complete syntheses are very important. The chemists dealing with natural compounds continually isolate and analyze the structures of natural compounds such as vitamins, hormones, antibiotics, etc. They always try everything they can possibly think of in order to reproduce artificially the often extremely complex natural substances in their laboratories. If the compound with which they begin such a synthesis is a natural compound, or a compound which can be quite easily synthesized from natural substances and already contains the basic structures of the natural compound which is to be synthesized, then we speak of a *partial synthesis*. If, on the other hand, the chemist begins with simple, chemically easily accessible compounds, then he is attempting a *complete* or *total synthesis*. The latter requires a far greater number of reactions than a partial synthesis because of the need to arrive at the desired natural substance over a series of intermediate compounds. Since only very rarely reactions have a one hundred per cent yield, the higher the number of successive reaction stages, the lower the amounts of the desired product. Exceedingly refined and clever techniques for the handling of minute amounts of substances have had to be developed. Total syntheses of complicated natural compounds require both creative thinking and planning and a high degree of experimental ingenuity and skill in order to realize the planned steps of the synthesis. The complete synthesis of vitamin B_{12}, which will be discussed in this chapter, is an example of such a process.

Chemical Reactions under Extreme Conditions

Modern chemical research shows a general interest in the study of chemical reactions under extreme conditions such as extremely high or low temperatures, or very high pressures. High temperature chemistry, cryogenics, or high pressure chemistry are all disciplines which have developed in the 20th century.

High temperature chemistry is concerned with the chemical reactions of substances at temperatures over 1,000 °C. Cryogenics on the other hand studies chemical reactions in the temperature range between 0 °C and as close to absolute zero (i.e. 0 Kelvin = −273.16 °C) as we can get. One of the aims of cryogenics is the synthesis of compounds which cannot exist at higher temperatures. Also, the use of very low temperatures sometimes allows the chemist to avoid the creation of undesirable side products in the synthesis of certain compounds. Moreover, cryogenic chemistry is important for fundamental research in that it helps to clarify certain chemical processes. For instance, at very cold temperatures it is possible to isolate very unstable ions or free radicals[1] with the help of so-called matrices and to subject them to physical and chemical examination. These matrices are solids such as frozen solvents or polymers. Unstable particles, which have a very short life-time, are trapped in these matrices. Their enforced immobility prevents further reactions from taking place. Since decreasing the temperature always lowers the rate of chemical reactions such conditions make it possible to study single stages of fast reaction sequences in slow motion, as it were.

Chemical reactions at high temperatures have been known and practically exploited for thousands of years. We need only think of the extraction and casting of metals, the manufacture and glazing of ceramics and the production of building materials such as bricks. A modern branch of chemistry in which high temperature reactions play an important role is *plasma chemistry*.

The term plasma refers to a gas which, beside atoms, molecules, and free electrons, also contains positive and negative ions in such quantities that the sum total effect is one of an apparent electric neutrality for the gas. If the plasma also contains substances which are engaged in chemical reactions with each other, then it also contains free radicals, the unstable decomposition products of these reacting substances. Since these radicals and ions display extremely "reactive tendencies", chemical reactions involving them tend to proceed at inordinate speeds (10^{-3} to 10^{-5} s). In plasma chemical syntheses it is often possible to proceed directly from the raw material to the desired end product. The procedural short-cuts achieved in this manner are considerable compared to traditional methods and result in a high space-time exploitation ratio. The disadvantage of such procedures lies in the fact that plasma chemical reactions often produce a series of undesirable side-products.

Plasmas for the initiation of chemical reactions are usually produced electrically by way of silent discharges, high frequency discharges, point discharges, electric arcs, or as plasma beams.

57

---

1 Free radicals are counted among the molecules in an extended sense of the word. They consist of several atoms and contain unpaired electrons,

e.g. $H-\overset{\displaystyle H}{\underset{\displaystyle H}{C}}\cdot \triangleq$ a methyl radical

---

In plasmas which are formed by silent discharges or in high frequency fields under low pressure (1.333 Pa to 13,330 Pa = 10^{-2} Torr to 10^2 Torr), collisions with electrons which have been accelerated in the electric field excite the gas molecules, ionize them, and/or split them into atoms and radicals. Because their mass is so minimal, the free electrons can only transmit a tiny fraction of the energy which they have picked up from the electrical field to the other particles in the gas with which they collide. The most salient characteristic for plasmas produced in this manner is the fact that the temperature of the electrons is very different from that of the ions and the neutral particles (molecules, radicals and atoms). Whereas the electrons have temperatures of ca. 5,000 to 200,000 K, the ions and neutral particles have only ca. 300–1,000 K. These plasmas, which display such an inequitable distribution of energy are called non-thermal plasmas. The causes for this unequal distribution of energy can be found in the fact that the free electrons, because of their minimal mass, can only pass on a small fraction of the energy which they have picked up from the electric field in each collision with an atom, a molecule, or a radical, and in the fact that the number of collisions at these given low pressures is very small.

At higher pressures ($\geqq 0.1$ MPa) the number of collisions between particles increases considerably. All particles acquire the same temperature. A thermal equilibrium is established and we speak of thermal plasmas. In such plasmas, the atoms, molecules, and ions have a much higher temperature (2,000–6,000 K) than in the non-thermal plasmas. They are thus ideally suited for the synthesis of substances whose formation requires a great deal of energy (endothermal reactions) and does not occur at all at low temperatures. Examples of thermal plasmas are electric arc plasmas and the plasma beams produced with the help of an electric arc. Plasma beams in plasma torches are usually produced by blowing carrier gases such as hydrogen or nitrogen through an electric arc. To the thus formed plasma beam, suitable reaction gases such as methane or ethane are added. The carrier gas, which functions as the energy transmitter, is frequently also the reaction partner at one and the same time. A hydrogen plasma beam, for instance, that has had the reaction gas methane admixed to it, will contain (at a couple of thousand K) particles such as H, H_2, CH, CH_2, CH_3, CH_4, C_2H_2, C_2H_4, C_2H, C_3H, C, C_2, and others. Depending upon the reaction time of the chemical reactions which are taking place in the plasma beam, there is a maximum concentration of specific components at particular points in the plasma beam. If, for instance, the industrially so valuable ethyne (C_2H_2) is to be obtained from the plasma beam, then the plasma beam must be cooled rapidly at the point of maximum ethyne concentration in order to prevent a further thermal decomposition of the ethyne in the plasma. This cooling is generally achieved by guiding the plasma beam over a cooled metal surface or by injecting a liquid into the plasma beam.

The extraction of ethyne and ethene from the methane in natural gas, from the hydrocarbon fractions of crude oil or heavy oils by means of hydrogen plasma beams has already achieved industrial importance. Today, research and development in the field of thermal plasmas is directed, among others, towards the economical production of such basic compounds in industrial synthesis chemistry as ethyne, dicyanogen, cyanide, and nitrous oxides, towards the extraction of high smelting oxides, nitrides, carbides and borides, towards the production of steel and aluminium, and towards the gasification of coal into hydrocarbons with the aid of hydrogen plasmas.

Very recently, a number of syntheses for organic compounds have been developed that employ non-thermal plasmas. These synthesis methods are frequently distinguished by the fact that it is possible to eliminate a number of reaction stages in comparison to conventional methods, and that in a single reaction process (for instance) it is possible to achieve a high rate of return of the desired substance. Plasma reactions are used both in the production and the removal (plasma etching) of thin layers of inorganic or organic compounds. This latter possibility is especially exploited by the micro-electronics industry.

fluorenon   silent discharge   biphenylene   + CO

The plasma-chemical transformation of fluorenon into biphenylene with a 90 % yield.

In the field of non-thermal plasmas, the present areas of challenge are in such spheres as the more economical production of ozone, which is already being produced with a plasma-chemical process, the development of economical syntheses of such inorganic basic chemicals as hydrogen peroxide and hydrazine, the elaboration of more syntheses for organic compounds, the production as well as the machining of thin sheets of polymers, the production of substances with the highest possible purity, and the improvement of textiles.

The above-mentioned examples show the intensive development in plasma chemistry taking place at the present time. Basic research is primarily concerned with the examination of the problem of energy transference in the plasma and in the researching of the fundamental chemical reactions.

As well as high temperatures, high pressure can also make chemical reaction possible and/or alter the characteristics of substances as well as influence the velocity and direction of many chemical reactions. The branch of chemistry which concerns itself with the changes in a compound's properties and chemical reactions under high pressure is known as high pressure chemistry, a term which generally indicates pressure conditions of above 10 MPa[1].

The high pressure synthesis of ammonia (see also Chap. V) is a graphic example of how the chemical equilibrium of ammonia, which makes its synthesis uneconomical at normal pressure, is altered by a higher pressure (reaction pressure approx. 20 MPa) so as to make possible an economically viable synthesis of ammonia. A pressure increase always favours that chemical reaction which reduces the volume of the substance upon which the pressure is exerted. That is the kind of reaction involved in the formation of ammonia from nitrogen and hydrogen:

$$3H_2 + N_2 \rightleftharpoons 2NH_3$$

4 volumes $\xrightarrow{\text{decrease in volume}}$ 2 volumes

This particular synthesis is carried out under increased pressure on elements in a gaseous state. In recent decades, we have also managed to carry out successful syntheses with substances in a liquid or solid state. The high transformation of substances at high pressures is primarily associated with the name of the American physicist P. W. Bridgman who in 1914 successfully transformed white phosphorus into black phosphorus by heating it under a pressure of more than 1000 MPa. It was physicists like Bridgman who specified the theoretical and experimental parameters of high pressure chemistry.

Synthetic Diamonds

During the early and middle 1950s, methods for synthesizing diamonds from graphite were developed. All the methods were examples of the way in which high pressure can decisively alter the properties of a substance.

---

1 1 MPa = 10 at.

Diamond is the hardest substance that can be found in nature. It is this property of hardness that determines its use as a tool in, for instance, the manufacture of diamond drills and saws, extrusion dies in wire production, drill heads for deep drilling projects, bearings for watches and for other precision instruments. The hardness, refractivity and the chemical stability of diamond also make it into the most highly valued of gems. Only a very small proportion of the diamonds found in nature are cut and polished for use in jewellery since not all natural diamonds can meet the high quality requirements. Natural diamond has a serious disadvantage which, on the other hand, increases its value immensely: it is very rare. Chemically speaking, diamond is a form of the element carbon. There is yet another natural form of carbon, graphite[1], which is not at all rare. With graphite, the most important constituent of the common pencil lead, it is possible to write; hence its name—*graphium*—from the Latin for pencil. Although graphite is constructed of the same carbon atoms which make up diamond, it has completely different properties. Graphite is a soft substance with a grey-black colour and a metallic sheen; unlike diamond, it conducts electricity. It is these properties which have determined its widespread use in pencil leads, as an emollient or lubricant, and as a raw material in electrode manufature.

Just as it is possible to build extremely diverse structures from the same type of bricks, so it is possible to construct very different modifications of an element from the same type of atom. The different properties of diamond and graphite are directly due to differences in their crystal structures.

The graphite lattice shows carbon atoms arranged in layers consisting of interlocking honeycomb-like hexagons. The $C-C-C$ bonding angles in the hexagons equal $120°$ and the distance between two C-atoms is $0.141\,nm$. These hexagons extend into two dimensions which is why we describe graphite as a two-dimensional macromolecule. The distance between two hexagonal layers is $0.334\,nm$. Since the C-atoms in graphite are connected to only three other C-atoms by means of σ-bonds (the overlapping of sp^2 orbitals), only three of the four bonding electrons in carbon are occupied. The fourth electron remains relatively free, similar to the free electrons in the metallic bonds. This is the origin of graphite's metallic sheen and its electric conductivity. The layers of carbon atoms in graphite are not interconnected through bonding between atoms. Instead, they are held together by the rather weak van der Waals forces. Quite moderate forces are capable of dislocating these layers, which explains the soft-

1 A further carbon modification, which does not occur in nature, is carbin. This was synthesized by Soviet scientists (Sladkov and others) in the 1960s and represents the "one-dimensional modification" of carbon. There are two forms of carbin, the polyin form $[-(-C\equiv C-)_n]$ and the comulene form $[(>C=C<)_n]$.

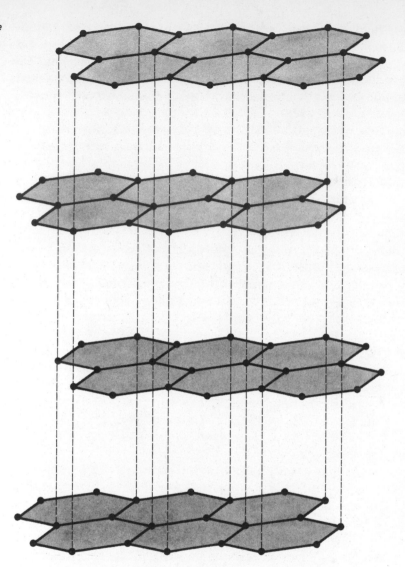

62

ness of graphite and its usefulness as an emollient and a lubricant. The different densities of diamond and graphite can also be explained by the different crystal structures. In the diamond lattice, the carbon molecules are much more tightly packed than in the graphite lattice. This tight packing of the atoms determines the higher density of diamond, 3.5 g/cm³, compared to 2.26 g/cm³ in graphite.

In diamond, every carbon atom is bonded with σ-bonds (the overlapping of sp³ orbitals) to four other carbon atoms in a tetrahedron structure. The distance between two carbon atoms is always 0.154 nm. In this way, a three-dimensional structure is formed with no layers that can be displaced and no free or loosely bound electrons. This completely different bond arrangement and spatial geometry results in the totally different properties of diamond compared with those of graphite.

Graphite occurs much more frequently in nature than does diamond; it can also be economically synthesized from coke. Thus, it was an odd dream of chemists to transform the available and common graphite into the rare and valuable diamond. Since the end of the 19th century there have been many unsuccessful attempts at such a synthesis. Since the carbon atoms are much more tightly packed in the diamond modification than in the graphite modification, the idea of using pressure in order to induce the transformation practically suggested itself. Theoretical estimates indicated that, at a temperature of 700°C, a minimum pressure of 3,500 MPa would have

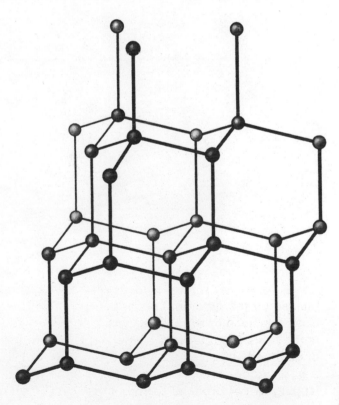

Diamond lattice

to be exerted upon the graphite in order to achieve the transformation into diamond. The problem of diamond synthesis was thus reduced to the technological problem of achieving and maintaining sufficiently high pressures. On the basis of the findings of high pressure physicists, it became possible to solve this technological problem. In 1954, the first reports of successful diamond syntheses were released in the U.S.A. and Sweden. In the U.S.A., it was Bundy et al. who first reported success. This research group worked with aliphatic hydrocarbons which were first decomposed into graphite. In the presence of catalysts (certain metals or alloys) this graphite was subjected to a pressure of 5,000 MPa and a temperature of 3,000 °C for 16 hours. It was possible in such reaction conditions to synthesize diamond crystals of up to 1.2 mm.

The close relationship between the process of diamond synthesis and economic interests can be seen in the fact that the successful research groups were employees in the laboratories of large industrial concerns (U.S.A.: General Electric Co.; Sweden: Almänna Svenska Elektriska Aktibolaget).

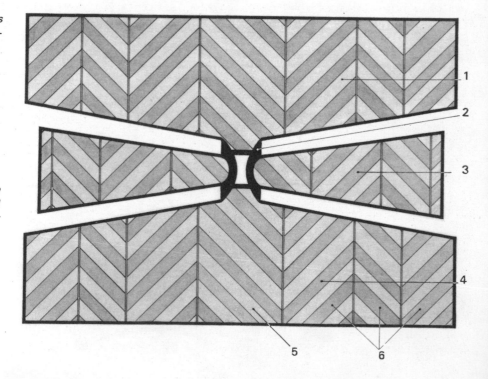

Belt apparatus for the production of synthetic diamond, after Hall. This belt apparatus produces the necessary pressure and temperature for diamond synthesis. The pressure on the graphite carbon is exerted by a press with the two stamps 1 and 4. The test chamber 2 is surrounded by a belt 3. The stamps and the belt consist of tungsten carbide 5 and steels 6. The test chamber 2 uses pyrophyllite as pressure transmitter, a silicate mineral which becomes plastic under high pressure. Pyrophyllite contains a heating element of metal or graphite for the production of the necessary reaction temperatures by means of electricity.
(After J. Liebertz, in: Naturwissenschaften 65, [1978], p. 503.)

With the high pressure apparatus developed in the U.S.A. at that time, it was possible to attain a pressure of 15,000 MPa, a temperature of up to 2,000 °C, and to maintain both for hours at a stretch. For instance, for an economic diamond synthesis the following reaction conditions were chosen: a temperature of 1,500 °C, a pressure of about 6,000 MPa, and nickel as a catalyst. Today, the pressures commonly used in diamond synthesis range between 5,000 MPa and 40,000 MPa and the temperatures between 500 °C and 4,000 °C.

The first syntheses produced diamonds as tiny crystals of only about 0.1 mm in size. Such "diamond dust" is largely used in abrasives, including those which are used for grinding and polishing diamonds themselves. Later, it became possible to produce larger crystals, suitable for jewellery.

The production and control of high pressures was not only important for diamond synthesis. A further interesting development in high pressure synthesis chemistry is the transformation of the hexagonal crystalline boron nitride (BN) into Borazon. Borazon is a cubic crystalline modification of boron nitride; it is even harder than diamond. Should an economically sound synthesis of Borazon become feasible, it would replace diamond as the premier material for hardness in industrial applications.

Syntheses at high and extremely high pressures require a considerable investment both in apparatus and in buildings—and are thus expensive. In the Soviet Union, for instance, very high pressure installations are of the size of a ten-storey building. Of course, the aim of an economic synthesis is always to produce highly useful substances at the lowest possible cost. This also applies to diamond synthesis and has recently resulted in a novel approach to the entire problem. This new approach is based on the theory of crystal growth. We all know the effect of introducing a few sugar crystals into a super-saturated sugar solution. The added crystals function as a sort of "seeds"; sugar molecules from the solution condense on the surface of these crystals which thereupon grow and form large sugar crystals. Analogous to this process, it has proved possible for tiny diamond crystals to "grow" in a super-saturated carbon atmosphere. Experimentally, this has been done in the following manner: Tiny diamond crystals were placed in a quartz tube. Then, at a pressure of 27 Pa and a temperature of 1,050 °C, methane (CH_4) was passed over these tiny crystals. In these conditions, methane decomposes into carbon atoms which, however, apart from contributing to the growth of the diamond crystals precipitate as graphite, a byproduct which causes impurities in the desired diamond product. Fortunately, there is a technique whereby this source of impurities can be removed. If the growth process is interrupted after a period of time and the reaction pipe is flushed with hydrogen at 5 MPa, the graphite is transformed into volatile hydrocarbons and thus removed. Diamond reacts very much more slowly with hydrogen than does graphite and is thus hardly affected by this procedure. The above process of diamond synthesis can be repeated in any number cycles consisting of the decomposition of methane and the removal of the graphite; the process can be optimized and automated, and we have hopes of producing considerably larger and cheaper diamonds with the help of this technique than with the high-pressure method.

In the early 1950s diamond syntheses formed the central concern of high-pressure synthesis. More recently, attention has been shifted to the synthesis of metallic hydrogen. The element hydrogen normally occurs in gaseous form, not surprisingly, since its melting point is 13.8 K = −259.4 °C and its boiling point is 20.4 K = −252.8 °C. In its solid state, normal hydrogen has a density of only 0.09 g/cm³. This natural hydrogen modification is constructed of diatomic molecules and does not conduct electricity. Very high pressures indeed are required to alter the bonding patterns in such a substance so that it acquires metallic properties. Successes in the area of the synthesis of metallic hydrogen have been achieved primarily by the Institute for the Physics of High Pressure of the Academy of Sciences in the Soviet Union. Researchers at this institute recently managed to change hydrogen into metallic hydrogen at a static pressure of more than 100,000 MPa. Under such conditions, metallic hydrogen does exist.

The aim of this research, however, is more far-reaching. It is to synthesize a metallic modification of hydrogen which would be *stable* at normal pressures and up to some hundreds of degrees Kelvin. That such a thing is possible has reliable theoretical support. The great interest in metallic hydrogen is largely due to the fact that such a substance should have superconductive properties, i.e. it would transport electricity without resistance or with only negligible resistance. Since all known superconductors acquire this property only at temperatures below 20 K, a super conductor which would still be stable at room temperature (300 K) would be invaluable for electrical engineering and electronics.

The incredible material expense involved in the research on the synthesis of metallic hydrogen becomes evident if we consider that the above mentioned institute has a high-pressure press with a mass of 5,000 tons, with which static pressures of up to 300,000 MPa can be produced. Pressures of this magnitude are thought to exist in the earth's core, and the equipment is used in testing models of chemical processes which may be occurring there.

Noble Gas Compounds

In the early 1960s one of the most exciting fields in chemical synthesis was the synthesis of noble gas compounds. The group of noble gases consists of the elements helium (He), neon (Ne), argon (Ar), krypton (Kr), xenon (Xe) and radon (Rn). At the time, it was virtually chemical dogma that noble gases cannot enter into chemical combinations with other elements. This was well supported by all the failed attempts at producing noble gas compounds as well as by the theoretical notions concerning the stability of the outer electron shells of noble gas atoms.

True, it had been possible to produce the so-called hydrates of the noble gases with the approximate constitution of $X \times 6H_2O$ (X = Ar, Kr, Xe), but these hydrates are not chemical compounds in the strict sense. They display neither the ionic nor the covalent bonds. The noble gas hydrates belong to the group of the clathrates, the so-called interstitial species. In the hydrates of noble gases, the water molecules form a crystal lattice in such a way that 46 water molecules enclose 8 "sites" on which noble gas atoms can be accommodated. When this occurs, the numerical ratio of noble gas atoms to water molecules is 8 : 46 or 1 : 5.75 which explains the "approximate" constitution of the noble gas hydrates, namely $X \times 6H_2O$. The noble gas atoms in the hydrates are locked into the "prisons" of the crystal lattice interstices but valence electrons[1] remain uninvolved; chemically speaking, nothing has happened. Noble gases also form clathrates with other compounds such as toluene, phenol, and hydroquinone.

In 1933, Linus Pauling[2] suggested that it should be possible to induce xenon and krypton to react with each other with the help of the strongly electrophilic fluorine. However, experiments by other researchers in this direction were not successful. The noble gases remained aloof and noble. Then, in 1962, more than sixty years after the discovery of the noble gases (1894–1900) Neil Bartlett[3] succeeded in breaking the "sound barrier" of the non-reactivity of the noble gases. He managed to bond xenon with the help of the extremely

1 Valence electrons are the electrons in the outer quantum shell of an atom which are available for bond formation.
2 Pauling, Linus, born 1901, American chemist. Nobel Prize for Chemistry in 1954, and Nobel Peace Prize in 1963.
3 Bartlett, Neil, born 1932, Canadian chemist, working in the U.S.A. since 1966.

electrophilic platinum (VI) fluoride, PtF_6, into an orange-red solid xenonfluoroplatinate with the chemical constitution $Xe[PtF_6]n(1 < n < 2)$. This was the very first noble gas compound to be synthesized. In this compound, xenon has a partly positive charge and the PtF_6-group a negative charge. In 1962, further syntheses were reported in the U.S.A., the Federal Republic of Germany, and Yugoslavia. These were xenon (II) fluoride XeF_2, xenon (IV) fluoride XeF_4, and xenon (VI) fluoride XeF_6. XeF_2 was achieved by irradiating a xenon fluorine mixture with ultra violet light; XeF_4 was the result of a reaction between xenon and fluorine at temperatures of up to 400 °C; XeF_6 was synthesized by heating the two elements at high pressures. The reaction vessels used in these syntheses are usually made of nickel since this metal is not attacked by the exceedingly reactive fluorine. Where the reaction mixture needs to be irradiated with ultra violet light, the reaction vessels are equipped with sapphire windows.

Xenon fluorides are colourless crystalline compounds which are relatively stable at room temperature provided that they are protected from moisture. The crystal lattices of these fluorides are made up of molecules. This is also the reason for the low melting points of these compounds (XeF_2 about 140 °C; XeF_4 about 114 °C, and XeF_6 47.7 °C). Some details of bonding patterns in the noble gas compounds are still being debated but in principle all questions can be answered by existing theories of chemical bonding.

After the successful syntheses of 1962, there rapidly developed a "noble gas chemistry". In the following we give a selection of the presently known noble gas compounds:

XeF_2	$XeCl_2$	$Xe[PtF_6]n$	KrF_2
XeF_4	XeO_3	$Xe[OTeF_5]_6$	RnF_x
XeF_6	Ba_3XeO_3	$CsXeF_7$	

A particularly dangerous substance among the noble gas compounds is the xenon trioxide, XeO_3, which explodes very easily. The above selection shows no compounds for helium, neon, or argon, and not without reason. Up to now, no stable compound with these noble gases has been achieved. The reason for this is that the valence electrons of these elements are even more firmly bound than those of krypton, xenon, and radon. If the noble gases are arranged in the order of their increasing reactive capacity, then the order is He, Ne, Ar, Kr, Xe, and Rn. This implies that chemical compounds with radon as the noble gas component should be the easiest to synthesize. However, to date, the only radon compounds have been fluorides whose exact constitution remains unknown (that is why there is an x in the formula RnF_x). The solution to this puzzle can be found in the radioactivity of radon. The most stable radon isotope, one with the mass number 222, has a half life of only 3.8 days. To experiment with an element that literally falls apart in one's hands is no easy matter. Furthermore, radon is a rare element, only available in very small amounts, so that it is necessary to work with minute amounts of it. The existence of the above mentioned radon fluorides could be established indirectly because the radioactivity on the walls of the reaction vessel increased when fluorine was mixed with radon. To be more specific, the formerly even distribution of radon radioactivity over the surface of the reaction vessel was cancelled. The explanation for this effect is that radon reacted with the fluorine to form a fluoride which was deposited on the walls of the reaction vessel.

The synthesis and structural analysis of noble gas compounds is an extremely important piece of fundamental chemical research. It was achieved through the combined use of traditional and modern experimental, structural and analytical methods. At present, there are already speculations as to possible uses for the noble gas compounds. Among them are suggestions such as the use of the fluorides of the noble gases for the separation of the radioactive isotopes of xenon and krypton during the regeneration of nuclear fuels; the use of solid noble gas fluorides as handy fluoridating agents and the use of xenon trioxide as an explosive.

Molecular Architecture: The Games Chemists Play

One branch of chemistry which consistently challenges chemists to come up with original and ingenious achievements is the synthesis of interesting, aesthetically pleasing, or unusual ring systems. Structures 1 to 12 illustrate what has been achieved in this area up to the present time.

Among these ring systems we find representatives of the ring-form alkanes, the cycloalkanes, and combinations of these cycloalkanes into exceedingly complicated structures. It is possible to recognize three, four and five rings on the quadricyclane. The simplest representatives of the cycloalkanes are cyclopropane (Formula 1) and

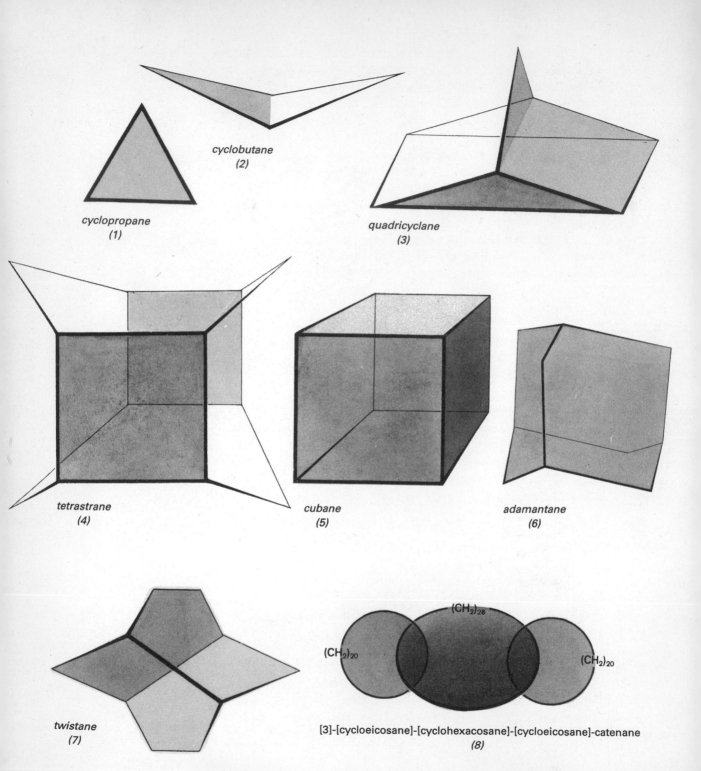

cyclopropane
(1)

cyclobutane
(2)

quadricyclane
(3)

tetrastrane
(4)

cubane
(5)

adamantane
(6)

twistane
(7)

$(CH_2)_{26}$

$(CH_2)_{20}$

$(CH_2)_{20}$

[3]-[cycloeicosane]-[cyclohexacosane]-[cycloeicosane]-catenane
(8)

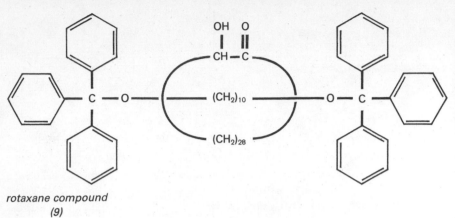

rotaxane compound
(9)

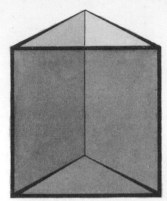

prismane
(10)

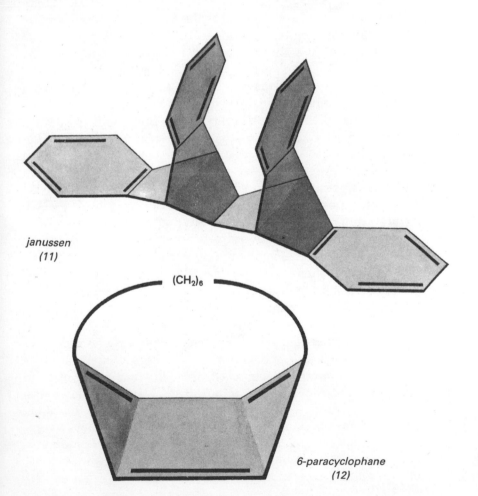

janussen
(11)

6-paracyclophane
(12)

Uncommon ring systems

The corner dots in the structural formulae symbolize carbon atoms which have bonded the number of hydrogen atoms (not shown) corresponding to their fourfold valence.

cyclobutane (Formula 2). The corresponding open-chain compounds are the alkanes propane and butane.

In these n-alkanes, the C–C–C angle of valence is 112.7°. In order to transform propane into cyclopropane, a closing of the ring between the C-atoms 1 and 3 must be induced. During this process, a distortion of the original bond angles takes place, as can be seen in the triangle formula of cyclopropane (p. 70). This distortion leads to stress in the molecule, i.e. cyclopropane is much richer in energy than propane. The process can be compared to the bending of a tensile rod.

The strain created in ring compounds through this distortion of the angles of valence relative to the corresponding open chain compounds is known as *"I-strain"* (internal strain). There are other forms of strain which can contribute to the overall strain of the structure. The more strain in a ring molecule, the more energy it has and the more unstable it is. Clearly these molecular architectural games of the chemists are also a means for establishing the limits of the stability of compounds.

The synthesis of unusual rings led to the development of new and original synthetic methods. Examples of such methods are the syntheses of the catenanes (Formula 8) and the rotaxanes (Formula 9). The catenanes are compounds whose rings interlock, like the rings of a chain (Latin: *catena* = chain). The first catenanes were synthesized in the early 1960s mainly by G. Schill of the Federal Republic of Germany. The unusual feature of these compounds is that the rings are not chemically bonded to one another. How, then, are the rings interlinked? The diagram on page 73 shows the principle of synthesis of a catenane consisting of three interlocked rings.

The starting compounds are A and B, the so-called *ansa compounds. Ansa* is the Latin word for handle, and we are here concerned with "two-handled" compounds. In our case, the "handles" consist of a series of CH_2-groups. In positions 1 and 3 of the aromatic rings there are open links, namely further CH_2-groups with reactive terminal groups X and Y. The coupling of the A and B molecules involves a separation of the reactive terminal groups and results in the compound C. The final step of the synthesis for a catenane consists in the splitting up of the bonds indicated by the letter a.

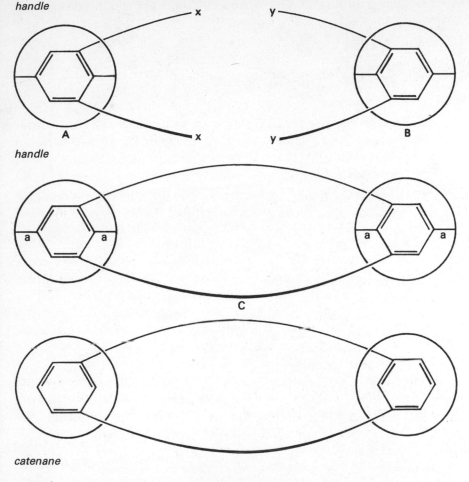

handle

handle

catenane

Rotaxanes, like the catenanes, are built of molecules which are not chemically bonded. In a very simplified form, the rotaxane compound could be diagrammed as demonstrated on p. 74.

The comparison with a wheel (Latin: *rota* = wheel) which is threaded by a spindle or axle clearly suggests itself. The ends of the spindle consist of groups of such bulk that the "wheel" cannot fall off. In chemical terms the spindle may be built from a series of methylene groups while the ends of the spindle contain three benzene rings each.

When first synthesized, the catenane and rotaxane structures seemed to be the incarnation of some rather bizarre ideas of eccentric chemists. This made the reports that compounds based on such structural

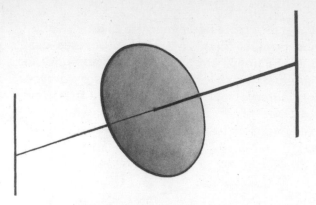

principles occur in nature even more sensational. The compounds in question are deoxyribonucleic acids, the DNA (see Chapter IV) and are central to all living matter.

The Total Synthesis of Vitamin B_{12}

A distinguished chapter in organic synthesis chemistry in the 20th century was written on successful completion of total synthesis of vitamin B_{12}. Vitamins are organic substances which are absolutely essential for the health and proper functioning of the human body and which must be ingested with food or otherwise introduced into the organism. Vitamin deficiency leads to syndromes such as beriberi, scurvy, rickets, pernicious anaemia, etc. A partial list of vitamins includes vitamin A, vitamins B_1, B_2, B_{12}, and vitamins C and D, all among the most common.

The history of the discovery of vitamin B_{12} is closely tied to the struggle against pernicious anaemia, a severe blood disease. This disease induces pathological changes and a decrease in the number of the erythrocytes of a patient. If the condition is not treated, it will result in death. Before 1926, pernicious anaemia could not be cured. In that year, the American physicians Minot and Murphy managed to cure pernicious anaemia by prescribing a diet of raw beef liver. The results were amazing. The conclusion is obvious: beef liver must contain a chemical compound which effects this cure. The hunt for this substance was on at once. Research began immediately, but it took 22 years until this anaemia curing substance, vitamin B_{12}, was finally isolated. The main reason for the length of the search was the fact that vitamin B_{12} is present only in minute amounts in the raw beef liver. At first, it could only be given in the form of liver extracts as part of a pernicious anaemia therapy. Today, B_{12} is harvested as a side product of the microbial processes used in the manufacture of the antibiotics streptomycin and aureomycin. In the presence of tiny amounts of cobalt the moulds *Streptomyces griseus* and *Streptomyces aureofaciens* produce not only the above mentioned antibiotics but also vitamin B_{12}. It is also possible to produce vitamin B_{12} relatively economically by using the so-called propioni bacteria. In its solid state, vitamin B_{12} comes in the form of dark red crystals which will absorb up to 12% water when exposed to air. It probably does not exist in nature, and is produced through the methods chosen to isolate it from the naturally extant co-enzyme B_{12}, the actual substance responsible for a pernicious anaemia cure. If vitamin B_{12} is ingested, the animal or human organism easily derives the co-enzyme from it. After vitamin B_{12} had been isolated, chemists embarked on an intensive research into the structure of the mysterious red crystals. It was a painstaking and slow business. Research teams headed by Lester Smith and Sir Alexander (later Lord) Todd[1] in Great Britain and by Folkers in the U.S.A. did the "chemical" analysis. The physics contribution to the clarification of the vitamin B_{12} structure was made primarily by a team of crystallographers headed by the British scientist Dorothy Crowfoot-Hodgkin[2]. They did an X-ray diffraction analysis of the crystals. By 1957, the complicated structure of the red crystals of vitamin B_{12}, a compound with the summary formula $C_{63}H_{88}CoN_{14}O_{14}P$ and a relative molecular weight[3] of 1355.42, had been established.

An important constituent of vitamin B_{12} is a nucleotide-like structure, called 5,6-dimethyl-benzimidazol-ribonucleotide (see p. 140). However, the most unusual structural group of the vitamin is the corrin ring system with its four pyrrole rings on a single plane. This ring system is arranged around a cobalt ion in a six-fold co-ordination. Four co-ordinations are occupied by the four pyrrole rings. A fifth is bound by the 5,6-dimethyl-benzimidazol-ribonucleotide residue. The broken line between the cobalt ion and this residue indicates that this grouping lies below the plane of the ring. The sixth co-ordination is occupied by a cyano group $(-CN)$ located above the plane of the ring. We have before us the actual vitamin B_{12}. If, instead, the sixth co-ordination is occupied by a 5'-deoxyadenosyl residue, it is the actual biologically effective B_{12} co-enzyme.

After the structure of vitamin B_{12} had been established, the synthesis of this complicated low molecular weight compound posed a considerable challenge to chemists. As early as 1960, W. Friedrich et al.

Max von Laue, German physicist, 1879–1960, received the Nobel Prize for Physics in 1914 for discovering the X-ray diffraction in crystals, which became the basis of the X-ray structural analysis.

1 Todd, Alexander Robertus (Lord Todd), born 1907, British chemist. Nobel Prize for Chemistry in 1977.
2 Hodgkin, Dorothy Mary Crowfoot, born 1910, British chemist. Nobel Prize for Chemistry in 1964.
3 Relative molecular weight of a compound is the sum of the relative atomic weight (see Chapter I) of the atoms which make up the molecule.

Corrin-ring system

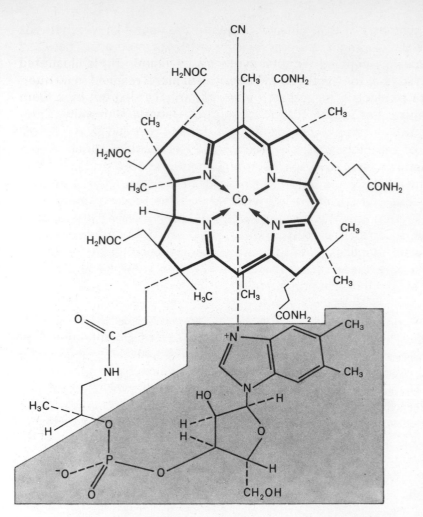

5,6-dimethyl-benzimidazol-ribonucleotide

were able to demonstrate that B_{12} can be synthesized from the so-called cobyric acid. The cobyric acid for this partial synthesis was obtained by first decomposing some vitamin B_{12}.

In 1962, a method was found for synthesizing the biologically effective co-enzyme B_{12} from the vitamin B_{12}, but this did not advance the search for the essential steps in conducting a total synthesis of vitamin B_{12}.

The main problem was posed by the synthesis of the cobyric acid, a compound which already contains the principal structural elements of vitamin B_{12}. The basic structure of the cobyric acid is represented by the rings labelled A, B, C, D. With nine chiral centres, cobyric acid displays an extremely complex spatial conformation.

The synthesis of this molecule required the use of known methods and the development of stereo-specific synthetic methods. The effort and expense spent on the total synthesis of vitamin B_{12} is illustrated by the fact that the synthesis of cobyric acid alone resulted in 60 intermediate products. The research was primarily carried out by a team of chemists headed by R. B. Woodward of Harvard University, Cambridge, U.S.A. and another team in Switzerland, headed by A. Eschenmoser of the Eidgenössische Technische Hochschule, Zurich. About 100 chemists were involved. Work was begun in 1960 and continued for twelve long years before they achieved success. In the synthesis of the four-ring system of the cobyric acid the research teams pursued different paths up to the synthesis of an important intermediate product, a corrinoid cobalt complex. After that, their further synthesis steps were identical. The different strategies pursued by these two teams in attempting to synthesize the four-ring system of the cobyric acid is shown in the summary on page 78.

The total synthesis of vitamin B_{12} also represents a tremendous achievement on the part of theoretical organic chemistry. In relation to research done for the synthesis of the bi-cyclic component A–D of the cobyric acid, it was possible to derive generally valid "Rules for the Maintenance of Orbital Symmetry" which are of general validity. These rules, named after Woodward and Hoffmann, are as important for the structural and reaction thinking of organic chemistry as was the tetrahedron model for the development of conformation analysis.

The structure of cobyric acid.

5′-deoxy-adenosyl residue

The synthesis of vitamin B_{12} (from A. Eschenmoser, in: Naturwissenschaften *61, 517* [1974]).
ETH = Eidgenössische Technische Hochschule

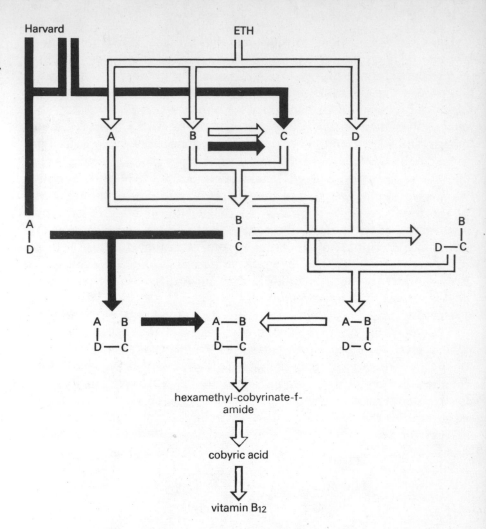

Harvard ETH

A B C D

A D

B C

D C

B C

D C

A B
D C

A B
D C

A B
D C

hexamethyl-cobyrinate-f-amide

cobyric acid

vitamin B_{12}

The resultant methods for the total syntheses of vitamin B_{12} are not, in fact, used for the manufacture of the vitamin. They are far too expensive compared with those methods of manufacture which use microbial processes. From the point of view of economics, moulds and bacteria are by far the best chemists in this area.

The importance of the total syntheses of vitamin B_{12} can be found in the concomitant discovery of the corrinoid natural substances. The research also contributed to our chemical knowledge concerning biologically important molecules and helped develop new methods and reactions in the field of synthetic organic chemistry. Its important theoretical contribution was the establishment of the Rules for the Maintenance of Orbital Symmetry.

Preceding page:
18 *The red crystals of vitamin B$_{12}$.*

19 *"The first X-ray picture of a crystal".*
In 1912, the physicists Laue, Friedrich, and Knipping discovered X-ray diffraction in crystals. This discovery led to the development of X-ray-spectroscopy and X-ray-structural analysis. The latter is a process whereby mathematical methods permit to reach conclusions as to the arrangement of atoms or ions in a crystal when they are applied to the direction and intensity of the diffracted X-rays. In order to register and measure these directions and intensities, various filming and counting methods are used. The photograph shows the historical first diffraction photograph which resulted when a copper sulphate crystal was irradiated with X-rays.

20 *Modern X-ray diffracto-meter.*

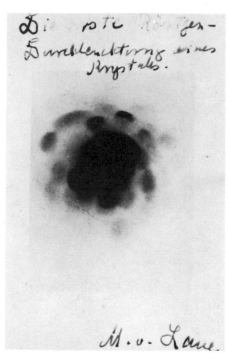

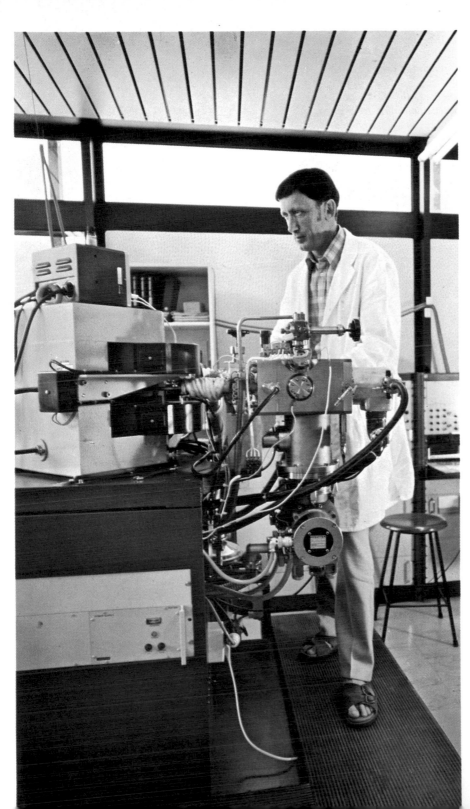

21 Mass-spectrometer linked to a computer for the processing of data.
Mass spectrometry is another modern physical method frequently used for the structural studies of chemical compounds. This method requires only very small amounts of the substance and can give information regarding the relative molecular weight and the structure of the analyzed compounds. Since natural compounds frequently can only be isolated in very tiny amounts, mass spectrometry is the method of choice for the structural analysis of such substances. The coupling of mass spectrometers to computers reduces the time necessary for the recording and evaluation of data in a mass spectrometric analysis.

22   Nitrogen plasma beam in irrotational emergence from a plasma torch.

23   Nitrogen plasma beam after the admixture of methane in turbulent emergence.

24   The apparatus for the synthesis of xenondifluoride from xenon and fluorine with the help of electrical discharges into the mixture. The transparent parts of the apparatus are made of quartz.

82   *25   Gel chromatography is used in the analysis of molecular weight distribution in polymers.*

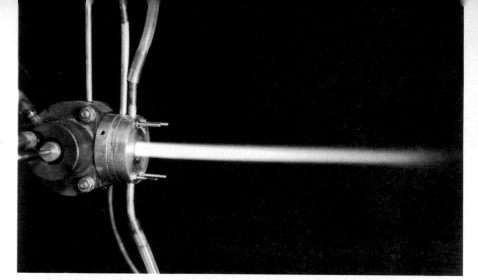

26 The use of plastic folios in agriculture brings about earlier harvests.

27 Polymerization vessel for the manufacture of Vinoflex, a polyvinyl chloride product.

28 The production of polyvinyl chloride tubing.

29 Abrasion test for the measurement of the wear and tear resistance of tyres. Tyres have been manufactured since ca. 1895. In 1920, it was discovered that the admixture of soot in the rubber increased the durability and abrasion resistance of the final product.

III.

The World
of Synthetic Macromolecules

For vitamin B_{12} (see Chapt. II) we gave a relative molecular weight of 1355. Compared with the relative molecular weight of water $(= 18)$ vitamin B_{12} seems a rather large molecule. However, there are molecules whose relative molecular weight is of the order of millions. These molecules are known as macromolecules (Greek: *makros* = large). The term, coined and introduced into chemistry by Staudinger[1] in the early 1920s, is now used to refer to molecules which consist of at least several hundreds of atoms and which are the result of covalent bonding. The lower limit for the relative molecular weight of a macromolecule is set at about 10,000. Staudinger's theory, namely that macromolecules do exist, was rejected by a number of leading chemists of his time. His colleague, Wieland[2], made the following comment:

"My dear friend, drop this idea of your macromolecules; organic molecules with a molecular weight exceeding 5,000 simply do not exist. If you purify your products, like rubber, then they will crystallize and you will see that they are low molecular substances after all."[3]

Macromolecules are also known as *polymers*. The term indicates that a macromolecule is made up of very many parts (Greek: *poly* = many, *meros* = part). These "parts", called *monomers*, represent specific "small" molecules. When many monomers combine, they arrange themselves in sequences, like pearls on a string. Thus "long chain molecules" come into existence. However, this idea of the structure of macromolecules met with a great deal of resistance from many chemists. They were simply not prepared to accept the possibility of such structures for natural organic polymers such as proteins, nucleic acids, cellulose, starch, and rubber. Some organic chemists

87

1 Staudinger, Hermann, 1881–1965, German chemist. Nobel Prize for Chemistry in 1953.
2 Wieland, Heinrich, 1877–1957, German chemist. Nobel Prize for Chemistry in 1927.
3 Staudinger, H.: *Arbeitserinnerungen* (Memoirs). Dr. Alfred Hüthig Verlag, Heidelberg, 1961, p. 79.

were so accustomed to the idea that only low molecular weight compounds exist, that they proposed ring formulae with low molecular weights for these polymers. For example, an eight ring structure which was held together by intermolecular forces was proposed for rubber. Despite of these efforts, evidence for the existence of macromolecules mounted. By the late 1920s the notion that cellulose, starch, rubber, and the proteins are composed of macromolecules was generally accepted. However, it was not until the 1950s that nucleic acids (see Chapt. IV) were shown to be constituted of the so-called nucleotides which form chains with hundreds of thousands of units.

The discoveries of macromolecular chemistry not only revealed the basic structure of natural polymers but also led to the synthesizing of polymers which do not occur naturally. We distinguish between three synthesis reactions which result in the formation of polymers: *polymerization, addition* and *condensation*. Polymerization refers to the stringing together of monomers which contain reactive double bonds or rings. Anywhere from a thousand to ten thousand monomers can be bonded together per second in this process.

Low- and High-Pressure Polyethylene

The synthesis of polyethylene is an example of polymerization of ethylene (= ethene) monomers. When two ethylene molecules react with one another, a low molecular weight compound is formed with a methyl group (CH_3) at one end and a carbon-carbon double bond at the other.

$$H_2C = CH_2 + H_2C = CH_2 \longrightarrow CH_3 - CH_2 - CH = CH_2$$

The continuation of this reaction produces long chains which can consist of up to 2,000 CH_2 basic units, CH_2 being the recurring structural element.

$$\ldots CH_2 - CH_2 - CH_2 - CH_2 - CH_2 - CH_2 \ldots$$

In terms of bonding, the following process occurs: the carbon atoms of the ethylene are promoted from sp^2 to the sp^3 hybrid state and can thus form an additional covalent bond. The terminal groups of the polyethylene molecules are identical to those formed by a bonding of two ethylene molecules. Such linear, non branching polyethylene chains are formed in the presence of the so-called Ziegler[1] catalysts at temperatures under 100 °C and pressures of 0.2 MPa (2 at) or less. This type of polyethylene is therefore described as a low-pressure polyethylene. The macromolecules formed through the polymerization of ethylene to polyethylene do not have a uniform size. They differ in the number of their monomer units and thus have different relative molecular weights. Another way of describing the same effect is to say that the macromolecules display different degrees of polymerization. The lack of uniformity in the length of the polymer chains is a general characteristic of synthetic macromolecular substances. Compounds with a low molecular weight consist of identical molecules. Since macromolecules differ in their relative molecular weight, it is not possible to give a definite molecular weight value but only the mean. Such a state of affairs clearly shows that research in the field of chemistry has to solve difficult tasks to produce synthetic polymers with a high degree of polymerization and with a uniform molecular weight. There is a further difference between polymers which are formed through interbranching of the polymer molecules. If ethylene is polymerized at temperatures of about 200 °C and at a high pressure of about 150 MPa, then the so-called high-pressure polyethylene is formed, containing structures of the following kind:

Hermann Staudinger, 1881–1965.

89

$$-CH_2-CH_2-\underset{\underset{CH_3}{\overset{|}{\underset{CH_2}{\overset{|}{}}}}}{CH}-CH_2-CH_2-\underset{\underset{CH_3}{\overset{|}{\underset{CH_2}{\overset{|}{\underset{CH_2}{\overset{|}{\underset{CH_2}{\overset{|}{}}}}}}}}}{CH}-$$

Molecular interbranching in high-pressure polyethylene.

In polyethylene, the macromolecular chains form an intertwined mass. The forces of attraction between chains is very small so that heat and pressure increase the mobility of the chains and produce a fluid-like consistency. This makes it possible to mould polyethylene.

1 Ziegler, Karl, 1898–1973, German chemist. Nobel Prize for Chemistry in 1963 together with Giulio Natta.

When the material cools, it becomes hard. The process of re-heating would again make the substance malleable so that further re-forming is possible. Polymers with these properties are called thermoplastic polymers. High- and low-pressure polyethylene can both be used for making pipes, for articles requiring injection moulding procedures, for insulating materials, and for various kinds of foils. Articles made from polyethylene can be found everywhere: in agriculture, in the chemical and electronic industries, in construction and in a wide range of consumer products.

The different structures in the high-pressure as compared to the low-pressure polyethylene determine their different properties. The chains in the low-pressure polyethylene do not interbranch and can thus fold themselves in a very orderly and regular manner, achieving a virtually crystalline order in the process. This polyethylene is, therefore, much more resistant to heat and distortion than the less ordered high-pressure polyethylene. In the latter, the lateral chains prevent an orderly folding of the polymer chains. Instead, we have a structure much like a loose ball of wool. This results in a higher degree of elasticity in the end product. Depending upon which property is most desirable in the final product, the manufacturer will choose the appropriate polyethylene as his raw material.

Polyvinyl Chloride

If we exchange a hydrogen atom in ethylene for a chlorine atom, then we have the monomer of polyvinyl chloride (PVC), today's most commonly manufactured kind of plastic. Although the vinyl chloride monomer was already discovered in 1835, it took 100 years before the mass production of its polymer became feasible.

Vinyl chloride (chlorethylene)

Polyvinyl chloride
n = degree of polymerization
= the average number of
monomers in a polymer
molecule.

Polyvinyl chloride is a solid, inflexible, horn-like substance. It is resistant to most concentrated acids and alkalis and physiologically harmless. PVC is made into pipes, tiles, foils, and moulded into an endless variety of forms; there is scarcely an area in our economy that does not use some form of it. One disadvantage of its hardness is that it is unsuitable for some purposes; products would be too brittle, too fragile. Two methods for making PVC softer and more flexible have been developed. One consists of adding certain compounds to the PVC which have a "softening" effect. Such softeners are for example tritolylphosphate (commonly known as tricresylphosphate) and bis (2-ethyl-hexyl)-phthalate. The molecules of these compounds squeeze between the PVC chains and reduce the forces of attraction between them. The resultant product is therefore more elastic and more pliable. This type of PVC, which usually contains about 30 % of the softener substance, is known as soft PVC. It is used in the construction industry as insulation and flooring, in the electronics for insulation, and further on in consumer goods such as, for example, boots and soles.

The second method for altering the properties of PVC consists of polymerizing vinyl chloride together with a similarly structured monomer, vinyl acetate. Such a process is known as co-polymerization and the resultant product is a co-polymer.

Polymers which consist of only a single kind of monomer unit are called homopolymers. Depending upon the number of different monomers present in a co-polymer, these are known as binary, tertiary,

Space-filling model of polyvinyl chloride.

co-polymers

The co-polymerization of vinyl chloride and vinyl acetate.

quaternary, etc. co-polymers. Since different monomers can be bonded to one another in different ways, a binary co-polymer already has four possible bonding patterns. For a simplified illustration of these patterns, we will assume two different monomers, A and B. The first possibility is that A and B will alternate arbitrarily, without rhyme or reason. This is the case in random co-polymers, whose structures are exceedingly irregular. In contrast to these, there are the alternating co-polymers in which A and B maintain a strict alternating sequence. A third type of co-polymer is the so-called "graft co-polymer". In these, A_n chains of the macromolecule have B chains of equal or unequal length "grafted" onto them. Fourthly, there are the block co-polymers in which large segments of A alternate with large segments of B.

Binary co-polymers:

Random co-polymers

$$-A-A-A-B-B-A-B-A-A-B-B-B-$$

Alternating co-polymers

$$-A-B-A-B-A-B-A-B-$$

Graft co-polymers

$$-A-A-A-A-A-A-A-A-$$

$$\begin{array}{ccc} | & & | \\ B & & B \\ | & & | \\ B & & B \\ | & \\ B & \end{array}$$

Block co-polymers

$$+A-A-A-A-A+B-B-B-B-B-B+$$

Polytetrafluorethylene

If all four hydrogen atoms in ethylene are replaced with fluorine atoms (F), then the resultant substance is tetrafluorethylene, a compound which can be polymerized to polytetrafluorethylene (PTFE). This polymer is particularly resistant to corrosion and temperature changes.

It is insoluble in all known solvents and retains most of its properties within a temperature range of $-268\,°C$ to $+260\,°C$. PTFE is used for

$$\begin{array}{cc} F & F \\ \diagdown & \diagup \\ C = C \\ \diagup & \diagdown \\ F & F \end{array}$$

$$-(CF_2 - CF_2)_n$$

tetrafluorethylene   *polytetrafluorethylene*

insulating, caulking, bearing and coating. It is mostly used in the aviation industry. As a coating, PTFE is also used in frying pans in order to make frying with little or no fat possible.

Polypropylene

A close relative of ethylene, propylene, can be polymerized into polypropylene.

$$_nCH_3 - CH = CH_2 \rightarrow -(CH - CH_2)_n$$
$$| $$
$$CH_3$$

propylene   *polypropylene*

Early attempts to polymerize propylene produced an amorphous and sticky polymeric substance. It was not until 1954 that Natta[1] managed to produce a crystalline polypropylene. Natta built on the experience of Ziegler who had managed to produce low-pressure polyethylenes with the help of catalysts in the early 1950s. Natta extended Ziegler's catalyst systems[2] and found the right catalyst for the synthesis of crystalline polypropylene. The reason for the existence of two polypropylenes can be found in a certain stereo-isomerism which can occur in such polymers and which is called *tacticity*. As a rule, stereo-isomerism can occur in any polymer chains which have $-CH_2-CHX-$ as a basic building block. In the case of polypropylene, $X = CH_3$. Polymer chains with such a sequence have three possible spatial arrangements. Hence, we have isotactic, syndiotactic, and atactic polymers.

The figure on page 94 above illustrates these stereochemical possibilities. The carbon chain (extended bonds) are located on the paper plane. The hydrogen atoms and the ligand X are located either in front of or behind this plane. The wedge-shaped bonds symbolize the bonds in front of the paper plane and the broken lines represent those behind it. In isotactic polymers, ligand X is regularly found on one side of the polymer chain, in front of the paper plane in our illustration. In the syndiotactic polymers, X alternates regularly on both sides of the polymer chain. In atactic polymers, X is distributed at random on either side of the polymer chain. The crystalline polypropylene which Natta polymerized is an isotactic polymer. The amorphous sticky polypropylene, on the other hand, is the atactic polymer.

1 Natta, Giulio, born 1903, Italian chemist. Nobel Prize for Chemistry in 1963 together with Karl Ziegler.
2 Ziegler catalysts are organometallic compounds. A classical example of this category of catalysts is the mixture of $Ti(C_2H_5)Cl_3$ and $Al(C_2H_5)_2Cl$.

Tacticity in polymers:

Isotactic polymers

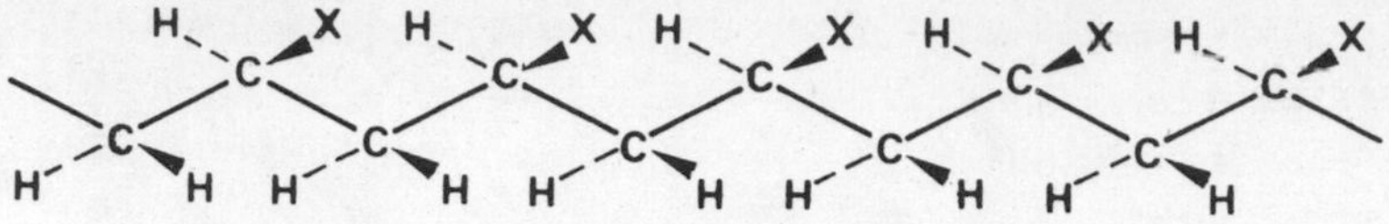

Syndiotactic polymers

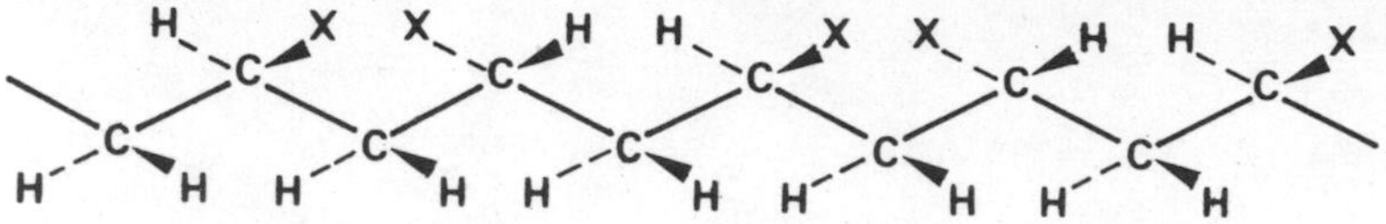

Atactic polymers

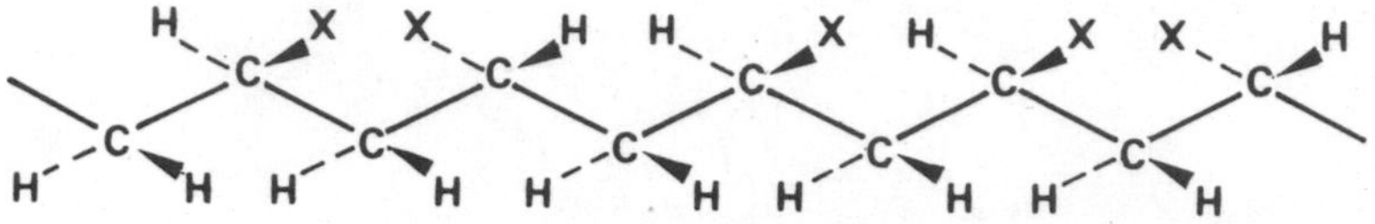

The different stereochemical structures of polymers are, in part, decisive for their properties. For this reason, modern polymer research is engaged in identifying these isomeric effects in order to use them in constructing synthetic polymers with specific properties.

Mass production of isotactic polypropylene was begun in 1956 in Italy. This polymer is extremely versatile. It can be manufactured into plasti-form bodies, spun into fibres, rolled into sheets and foils. The plasti-form bodies are primarily used in the motor industry and in the construction of washing machines, radios, and television sets. Polypropylene fibres are used in the production of technical textiles and in the manufacture of carpeting.

Natural and Synthetic Rubber

Natural rubber is a polymer, whose production in the early 1970s was three million tons annually. The most important rubber yielding plant is the indigenous Brazilian rubber tree *(Hevea brasiliensis)*. The Indians of South and Central America were caulking their boats with rubber and playing with rubber balls long before the Europeans colonized the Americas. In its natural state, rubber has the disadvantage of becoming rigid and brittle in the cold and soft and sticky in the heat. If the rubber is combined with sulphur at higher temperatures, this disadvantage disappears and the resultant rubber product is pliable and elastic. This process, known since 1839, is called *vulcanization*.

In 1905 Harries[1] discovered that the basic constituent of rubber is isoprene. It turned out that gutta-percha, which is also harvested from tropical trees, consists of the same monomers. However, the properties of gutta-percha are completely different properties from those of natural rubber: the material is horn-like, inelastic, and cannot be vulcanized.

isoprene (2-methyl-buta-1,3-diene)

The existence of these two polymers is again due to the stereochemical structure. In this case, we have a cis-trans-isomerism. This type of stereo-isomerism can be found in both ring-form compounds and in C=C double bond compounds. For an example of this isomerism, let us look at 1,2 dichloroethene.

cis-1,2-dichloroethene *trans-1,2-dichloroethene*

cis-trans isomerism in 1,2-dichloroethene

This compound exists in both a cis-form and a trans-form.

If we assume a plane of reference vertical to the double bond, then the two chlorine atoms in the trans-form of the compound are located on opposite sides of the plane of reference. In the cis-form of the compound, the chlorine atoms are both on the same side of the plane of reference.

Natural rubber is a cis-form of polyisoprene; gutta-percha is a trans-form of polyisoprene.

The above mentioned vulcanization process is of decisive importance for the elasticity of products made from natural rubber. In this process, the sulphur atoms react with the double bond in the rubber. Sulphur links are formed which interconnect the isoprene polymer chains with one another. This intermeshing causes the formation of large, three-dimensional molecules. If a force is exerted on this structure, a partial slippage of molecule chains past each other is all that the structure will permit. This explains why rubber tends to return to its original shape when the external force, be it stress or strain, is withdrawn.

1 Harries, Karl-Dietrich, 1866–1923, German chemist.

Vulcanization assures the elastic properties of rubber between $-60\,°C$ and $+100\,°C$. The main use of rubber is in the manufacture of tyres. Of course, these are not made of rubber only, but contain also a number of substances necessary to produce the properties that tyres must have. A typical tyre mixture consists of 100 parts natural rubber, 50 parts soot, 10 parts zinc oxide, 3 parts sulphur, 1 part vulcanization accelerator, 1 part age retardant, and 2 parts softener. As well as increasing its tensile strength the soot is responsible for the colour of the tyre.

96

Model representation of vulcanized natural rubber. The molecule chains are interconnected by sulphide bonds.

After the chemical structure of natural rubber had been established, the first attempts at synthesizing this isoprene compound were made in 1910. These did not at first succeed in the production of a pure *cis*-1,4-polyisoprene, i.e. natural rubber. Instead, the synthesized polymers were very different in length and were not even a good approximation of the properties of natural rubber. Also, isoprene, the raw material, was much too expensive to permit the manufacture of a synthetic rubber which could compete with the natural product. During World War I, the total lack of natural rubber in Germany led to the development of a synthetic rubber on the basis of 2,3-dimethyl-buta-1,3-diene, a compound which is very similar to isoprene:

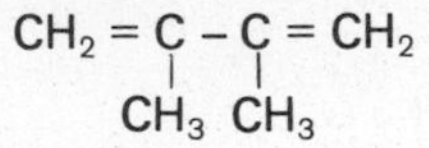

$$CH_2 = \underset{\underset{CH_3}{|}}{C} - \underset{\underset{CH_3}{|}}{C} = CH_2$$

2,3-dimethyl-buta-1,3-diene (methyl-isoprene)

However, this rubber substitute was purely an emergency product. Its production ceased when the war ended and natural rubber once more became available.

Today, the main raw material used for making synthetic rubber is buta-1,3-diene.

$$CH_2 = CH - CH = CH_2$$

buta-1,3-diene

Butadiene can be synthesized from acetylene (ethyne), which, in turn, is obtained from calcium carbide. The raw materials for calcium carbide are coal and lime. From these raw materials, Nazi Germany instituted mass production of butadiene rubber, an important strategic material in World War II. Butadiene can also be obtained from the crack gases when refining crude oil, from butane which occurs in natural gas and from ethanol.

Due to the presence of two double bonds in the molecule polymerization of butadiene can produce structurally different polymers. We distinguish between the 1,2-polymers and the 1,4-polymers. In the 1,2-polymers, only one of the double bonds opens up in order to provide the necessary reactive site for the polymerization. In the 1,4-polymers, the reactive sites are formed on the carbon atoms 1 and 4, with a double bond being formed between carbon atoms 2 and 3.

Four butadiene polymers of regular structure which are shown above, are possible if we consider that the cis-trans isomerism must be maintained in the 1,4-polymers, and tacticity in the 1,2-polymers. Industrially produced polymers consist of a mixture of 1,2- and 1,4-polymers. The properties of polybutadiene are improved through co-polymerization with styrene. This synthetic rubber, known as BUNA S, has been mass-produced since 1937 and is, on the whole, the equal of natural rubber. The name, BUNA S, is an abbreviation of the compounds involved: **bu**tadiene, sodium (German = **Na**trium) and styrene. Originally, sodium was used as a polymerization catalyst in the manufacture of polybutadiene.

The 1,2-polymerization of buta-1,3-diene.

The 1,4-polymerization of buta-1,3-diene.

$$n \ ^1CH_2{=}^2CH \quad \longrightarrow \quad -^1CH_2-^2CH{-}\left[^1CH_2-^2CH-\right]$$
$$\qquad\qquad\qquad\qquad ^3CH \qquad\qquad\qquad ^3CH \qquad\qquad CH$$
$$\qquad\qquad\qquad\qquad \parallel \qquad\qquad\qquad\qquad \parallel \qquad\qquad \parallel$$
$$\qquad\qquad\qquad\qquad ^4CH_2 \qquad\qquad\qquad ^4CH_2 \qquad\qquad CH_2 \Big]_{n-1}$$

$$n \ ^1CH_2{=}CH{-}CH{=}^4CH_2 \quad \longrightarrow \quad -^1CH_2-CH{=}CH-^4CH_2{-}\left[^1CH_2-CH{=}CH-^4CH_2\right]_{n-1}$$

Today, the term "synthetic rubber" applies to polymers many of which contain no isoprene at all. Their common property is their rubbery, elastic quality, for which reason they are also known as *elastomers*. Apart from the styrene-butadiene polymer, the synthetic rubbers include elastomers made of butyl-polymers, polychloroprene, poly-nitriles, polyurethane, and poly-olefins. In 1971, 6 million tons of synthetic rubber were produced, i.e. double the production of natural rubber.

Today, the synthesis of natural rubber as well as gutta-percha is possible with the help of the Ziegler-Natta catalysts. An industrial production of cis-1,4-polyisoprene, however, has not been developed since isoprene is not an economically competitive raw material. The most economic polymer in production is cis-1,4-polybutadiene. This "stereo-rubber" presently makes up 20 % of the synthetic rubber production. Its properties are very similar to those of natural rubber and some are actually superior. Cis-1,4-polybutadienes are primarily used in the production of tyre surfaces.

Synthetic Fibres

Polycondensation is a further method in the synthesis of macromolecules. The process is one in which bi- or multi-functional molecules are polymerized in a chemical reaction which involves the splitting out of a small molecule. We will examine polycondensation more closely, using the example of a reaction between adipic acid (butanedicarboxylic acid) and hexamethylenediamine. Both compounds are bifunctional, i.e. they both contain two reactive atom groups. In adipic acid, the reactive atoms are the two carboxyl groups $-COOH$, and in hexamethylenediamine, the two amine groups $-NH_2$.

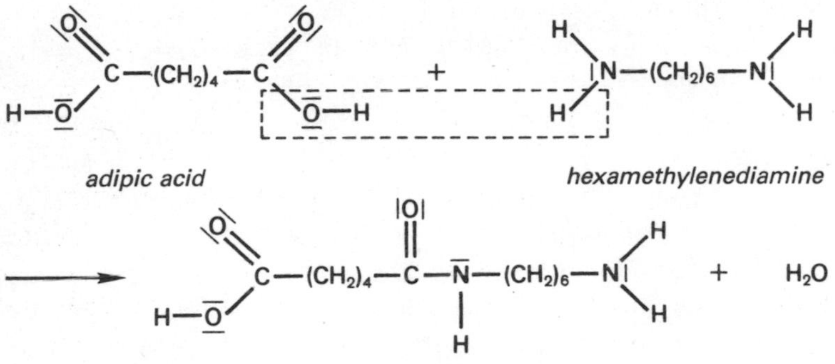

adipic acid   hexamethylenediamine

amide bond

The first step in the polycondensation of these compounds is the reaction of one carboxyl group of the adipic acid with an amine group of the hexamethylenediamine, splitting off the water molecule and forming an amide group. The resultant molecule is again bifunctional. The amine group of this molecule again reacts with the adipic acid and the carboxyl-group can now further react with the hexamethylenediamine. This is the synthesis method used in the synthesis of the so-called polyamides, i.e. compounds with many amide groups.

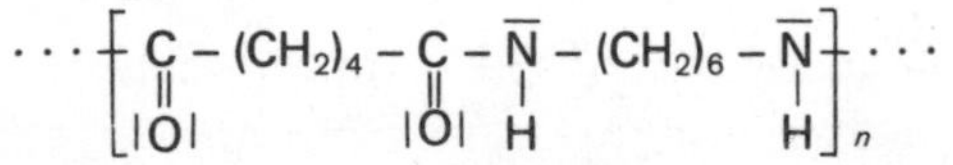

6,6-polyamide

Since this polyamide is constituted of two basic units with 6 carbon atoms each, it is called a 6,6-polyamide. It was first synthesized by

Carothers[1], who had begun synthesizing polyamides in 1929. Carothers was searching for a spinnable synthetic polymer for the production of textile fibres. He succeeded in producing a fibre from the 6,6-polyamides which has passed into the history of chemical fibres as *nylon*. In 1935 he took out a patent on his process for manufacturing nylon, the first synthetic fibre. Mass production of nylon began in 1939. It was carried out by the American chemical firm, DuPont, which had invested 27 million dollars in the development of nylon. The properties of polyamide fibres are, in some respects, an improvement over natural fibres. They are highly resistant to wear and tear and crinkle less. They are also more durable in relation to chemicals, mould and bacteria. Thus polyamide fibres are the material of choice in the manufacture of textiles.

Whereas nylon was first mass produced as a substitute for parachute silk, it rapidly gained importance through its use for sheer stockings, shirts, and blouses. The synthesis and mass production of nylon constituted an important advance for the chemical fibre industry. Up to that point, chemical fibre production had been almost exclusively limited to cellulose derivatives, i.e. fibres for which natural cellulose provided the raw material, i.e. the basic polymer.

Nylon was the first fully synthetic fibre which could be used by the textile industry. It was not, however, the very first technically usable synthetic fibre. As early as 1913, Klatte (in Germany) applied for a patent for the manufacture of fibres from polyvinyl chloride. Based on his idea, Schönburg and Hubert (also in Germany) developed the so-called PeCe-fibre at the beginning of the 1930s. This is made of polyvinyl chloride which has been re-chlorinated, by a further reaction of the polyvinyl chloride with chlorine resulting in carbon tetrachloride. The chlorine content of PVC increases from 57% to 64% during this process. PeCe fibres have a very low melting point which makes them unsuitable for the manufacture of general textiles. However, they are exceedingly resistant to chemicals and rotting and thus find their use primarily in industrial filters.

Perlon, the equivalent of nylon developed in Germany, is also a polyamide. In this case, the basic amide unit for the polymer is the ε-aminocaprolactam, a cyclic amide of the ε-amino caproic acid. In 1973, the world production of this compound amounted to more than 2.3 million tons. Since this lactam contains six carbon atoms, the polymerized version which constitutes the perlon is a 6-polyamide. Research by the German chemist Schlack (born in 1897) led to this extremely successful product for the chemical industry. The synthesis

1 Carothers, Wallace Hume, 1896–1937, American chemist, head of the Department of Organic Chemistry Research of E. J. DuPont de Nemours and Company as of 1928.

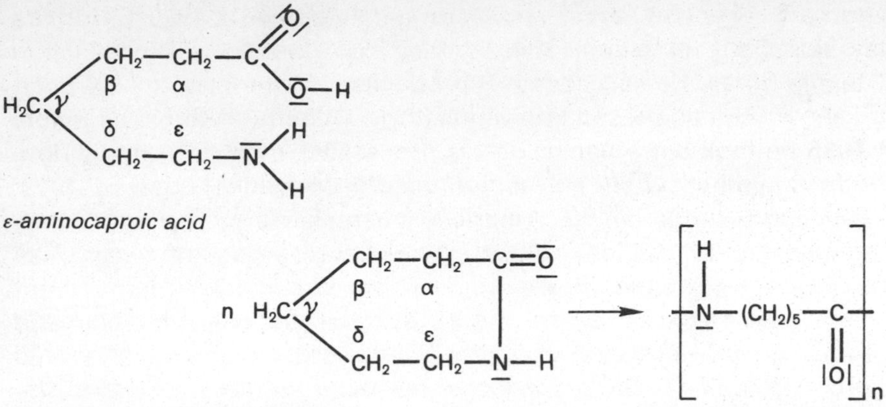

ε-aminocaproic acid

ε-aminocaprolactam (caprolactam)

6-polyamide (perlon)

of perlon was patented in 1938 and Schlack was able to close the last remaining gap in the patents on the polyamides, of which DuPont, at that time, already owned 130. This breach in the monopoly of the American concern was possible because Carothers had made a mistake in 1930. During his research, Carothers had erroneously concluded that ε-aminocaprolactam could not be polymerized, with or without catalysts. The work of Schlack corrected this error. With the help of ε-aminocaproic acid hydrochloride [HOOC $+CH_2+_5$ NH_3]$^+$ Cl$^-$ he managed to synthesize a 6-polyamide from the lactam which, "without further purification, could be spun to a highly elastic filament ... after cold drawing."[1]

A trial production of perlon was begun in Germany in 1939. Mass production did not start until 1943. The fields of the industrial and consumer use of this fibre are the same as for nylon.

Only a short time later, nylon and perlon were joined by a third synthetic fibre with especially wool-like properties. These were the *polyacrylonitrile fibres* (PAN fibres or simply the "acrylics") which were produced by polymerizing acrylonitrile (vinyl cyanide):

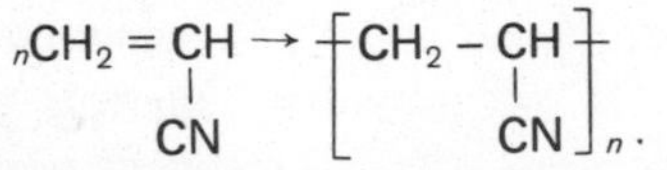

acrylonitrile   polyacrylonitrile

A patent application for this synthetic fibre was submitted in 1942 by H. Rein (Germany). Trial production was begun in 1943 in Germany

1 Schlack, P., *Pure and Applied Chemistry*, London, 15 (1967), p. 508.

and in 1944 in the U.S.A. The various trade names for fibres from this family of polymers are, among others, Orlon in the U.S.A., Dralon in the Federal Republic of Germany, and Wolpryla in the German Democratic Republic. Polyacrylonitrile fibres (PAN fibres) have an excellent resistance to light, heat, changing weather conditions, bacteria and moths. They are almost crinkle-free, fluffy and possess high insulating capacity. Apart from textiles, the PAN fibres are also used in the manufacture of tenting materials, filtering substances, and in the production of fishing nets.

The percentage of each of the various types of synthetic fibres produced in the world in 1976 is the following: the polyamide fibres account for 33%; and the polyacrylonitrile fibres for 20%. By far the largest amount however, 44%, is accounted for by the *polyesters,* a synthetic fibre that we have not as yet discussed. In the early 1970s, the production of polyesters exceeded that of the polyamides. The first place of the polyesters is justified by their superior properties compared with other textile fibres. Among these properties are good shape retention, high fluffiness, and an excellent crease and wrinkle resistance. Textiles made from polyester fibres are "wash and wear" and they dry quickly.

The synthesis of the polyesters.

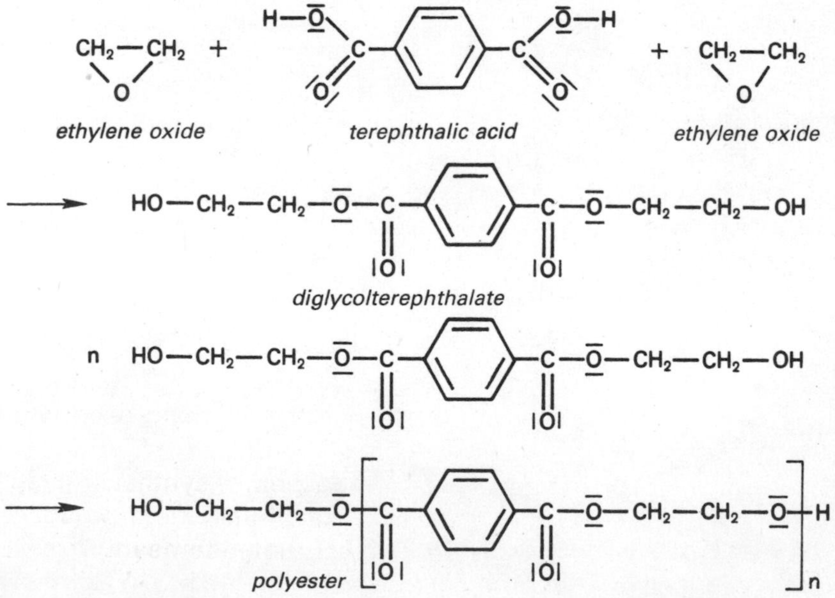

ethylene oxide   terephthalic acid   ethylene oxide

diglycolterephthalate

polyester

Polyesters are polymers with a recurrent ester group, $-\overset{\text{O}}{\underset{\text{||}}{\text{C}}}-\text{O}-$. The basic work involving polyesters was already done in the early 1930s by Carothers and Hill. In 1941, Whinfield and Dickson (Great Britain) took out the first patents involving the polymerization of polyester fibres, which they called *terylene*. Mass production of polyesters did not begin until the early 1950s. The most important polymer for the production of polyester fibres is polyethyleneterephthalate. This polymer is technically derived from a reaction of ethylene oxide (1,2-epoxyethane) with terephthalic acid. In this first step, the reaction product is diglycolterephthalate, which is then polycondensed in a second step.

Apart from the three main families of synthetic fibres, the polyamides, the polyesters, and the polyacrylonitriles, there are a large number of special fibres which are produced because of some particularly pronounced property such as high resistance to heat, chemical corrosion, or radiation. Among the temperature-resistant fibres, we must mention polymers which are derived from inorganic compounds. These hold a special place.

There are quite a few of them, made from metal, glass, ceramic, boron and carbon. Among the boron fibres are the *boron nitride fibres* which are chemically constituted of a single plane network of hexagonal rings of the following structure:

(Graphite displays a similar spatial structure; see p. 61). Boron nitride fibres can withstand temperatures of up to 3,000 °C, which makes them the most heat-resistant fibres known. They are used primarily for the reinforcing of plastics used in nuclear technology and in space exploration, e.g. for heat shields in rockets and spaceships. These fibres are also used in the manufacture of protective clothing used where there is danger of exposure to heat or radiation.

The immense development in the production of synthetic fibres is demonstrated by two numbers: in 1960 the annual world production was 700,000 tons, but already in 1975 it amounted to nearly 7 million tons.

Thermosetting Polymers

Among the synthetics which are primarily synthesized by means of the process of condensation are thermosetting polymers[1], substances which can be hardened by the influence of heat. These can be synthesized, for instance, from phenols and formaldehyde. At first, a variety of structurally different intermediate condensates are formed, which can still be melted. If these primary condensates are then reheated, with a possible further addition of formaldehyde, then a process of hardening ensues, i.e. a tight network of three-dimensional macromolecules is formed which can no longer be softened or melted without decomposing entirely. Such fully synthetic polycondensates of phenol or formaldehyde were first produced by Baekeland[2] in the U.S.A. in 1909, and achieved global prominence under their trade name "Bakelite".

phenol + formaldehyde

methylol compounds

104

The polycondensation of phenol with formaldehyde.

1. The reaction of phenol with formaldehyde, forming methylol compounds.

2. One possible reaction for the methylol compounds is that they will react with phenol, giving off water. With this reaction mechanism, the formation of three-dimensional internetted macromolecules is possible.

1 Polymeric organic compounds which are synthesized artificially or derived from natural compounds and which display plastic properties within a certain range of temperatures. They are characterized by their high degree of stability and minimal distortion under stress.
2 Baekeland, Leo Hendrik, 1863–1944, Belgian chemist, emigrated to the U.S.A. in 1889.

Polyurethanes

As the last time in this area of chemical endeavour, we will take a closer look at polyaddition, the third type of polymerization reaction for the synthesizing of macromolecules. In this chemical reaction, bi- or poly-functional molecules combine to form polymers without the splitting off of a low molecular compound. Also in contrast to simple polymerization, a shift of the hydrogen atoms takes place. The formation of the polyurethanes (PUR) is an example of addition. In this reaction, bi-functional hydroxyl compounds (alcohols, for instance) are brought to react with bi-functional isocyanates. The basic research in polyurethane chemistry was done by O. Bayer[1] after 1937. Mass production of polyurethanes was begun in the early 1950s. It is possible to synthesize both linear and branched polyurethanes. For instance, a linear, thermosetting polyurethane can be synthesized from buta-1,4-diol and hexamethylene-1,6-diisocyanate.

The synthesis of polyurethanes.

Polyurethanes of this kind have properties similar to those of the polyamides and are generally used in the production of fibres, bristles, felts, and foils.

If the linear PUR macromolecules are synthesized to form a network, then the result is an elastomer. Such polyurethane rubber or latex is remarkable for its durability and its resistance to oil. The material is especially useful in the manufacture of gaskets and membranes. This polyurethane is also used in the production of synthetic leathers.

Today, polyurethane is used primarily for the production of foamed materials such as insulating foam. The foaming or aeration of the polymer mass into a porous structure of low density can be achieved if carbon dioxide, for instance, is produced as a propellant gas during synthesis. We distinguish between hard and soft foamed materials.

1 Bayer, Otto, 1902–1978, German chemist.

The former have a loose lattice structure and are primarily used in the upholstery of furniture. Hard polyurethane foams on the other hand are highly latticed and thus have good shape-retaining properties. This makes them useful as structural elements for containers and furniture. Present world production of PUR foams exceeds one million tons annually, a figure which clearly demonstrates the economic importance of these products.

30 *cis-1,2-dichloroethene*

31 *trans-1,2-dichloroethene*

32   Manufacture of a 40,000-spinning nozzle out of a precious metal alloy. Spinning nozzles with 10,000 to 60,000 orifices and a diameter of 90 to 140 mm are employed, for instance, in the manufacture of acrylic fibres.

33   The spinning of polyacrylonitrile fibres in a solution spinning process. The moulding and stabilization of macromolecular linear polymers into fibrous formations is known as "spinning". For all spinning processes, the following is valid: the solution or melt of the polymer is extruded under pressure through (usually) metal pipes until it comes out through a multiple of spin-nozzles, where fibres are formed.

34 *High-voltage electron microscope.*
Electron microscopic examinations, in which sub-microscopic particles and structures are visible, are also used for the examination of polymers. With light microscopes, a maximum resolution is 200 nm, i.e. points 200 nm distant can still be seen as separate entities; anything smaller shows up as an undifferentiated lump. With high resolution electron microscopes, the resolution goes up to 0.3 nm which reaches into the atomic and molecular range.

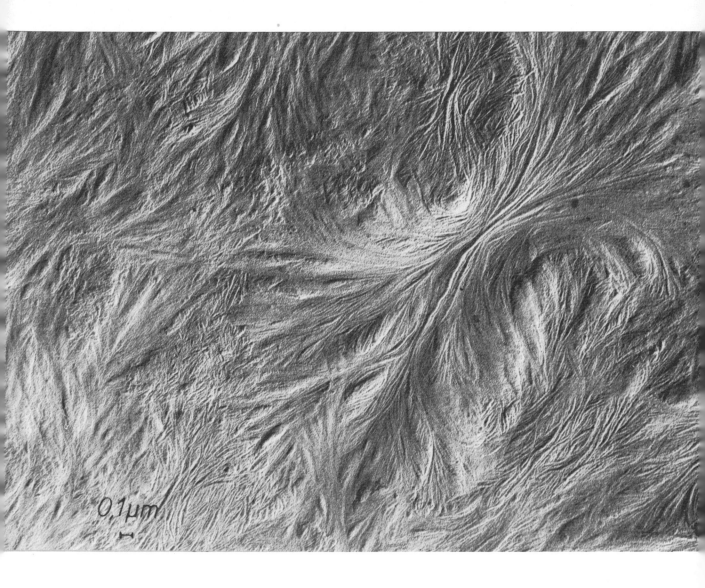

0.1μm

35 The electron-microscopic photograph of the spherolitic surface structures of a polyamide film.
If the molecular chains form a completely random aggregate, we speak of an amorphous material. A good comparison with such an amorphous state is to a bowl of spaghetti. If the macromolecules line up in a parallel, a state of extreme orderliness comparable with a crystalline order is achieved. This results in a high degree of strength of the material. With a regular chemical structure (as with low-pressure polyethylenes and polyamides) it is possible to obtain crystalline polymers from solutions or melts. With such processes crystallites can be formed which are known as spherolites and which are an important structural component of crystalline polymeric solids.

36 Equipment for the stretching of synthetic fibres. After spinning, the fibres are stretched to many times their original length. The procedure increases the linear orientation of the macromolecules and thus serves to strengthen the fibre.

37 Equipment for the dyeing of textiles.
The dyeing of textile fibres has been practised for thousands of years. In the past dyes were largely derived from natural sources, such as plants and animals. The natural organic dyes lost their importance in the 19th century when synthetic dyes were developed.

38   Synthesis of a rigid polyurethane foam.

a/b   A very few seconds after the reaction partners have been mixed (polyol and isocyanate components), the foaming action commences which results in an increased volume.

c   After completion of this process the foam solidifies.

114

IV.

The Molecular Foundations of Life

The Synthesis of Urea

On February 22, 1828, the 27-year old Friedrich Wöhler[1], a chemistry teacher at the Städtische Gewerbeschule in Berlin, wrote to his friend and teacher, the Swedish chemist Berzelius[2], in Stockholm:

"Dear Master!
...I cannot, so to speak, retain my chemical water and must tell you that I can make urea without using the kidneys of any animal, be it man or dog. Cyanic acid ammonia is urea ... I obtained the presumed cyanic acid ammonia quite easily by treating cyanic acid lead with caustic ammonia. It can also be obtained by using cyanic acid silver and ammonium chloride. I have synthesized a sizeable amount of it, nicely crystallized... This artificial synthesis of urea, can it be seen as an example of the derivation of an organic substance from inorganic matter? It is perfectly true that the making of cyanic acid (and of ammonia as well) always requires organic raw material, and a nature philosopher might say that in animal carbon as well as in the cyanide compounds derived therefrom the organic essence has not as yet been dissipated and therefore organic matter can always be produced from it.

Yours
Wöhler."

1 Wöhler, Friedrich, 1800–1882; a truly distinguished German chemist. Among other things, he discovered the element aluminium in continuation of Oersted's work towards the isolation of this element.
2 Berzelius, Jöns Jakob, 1779–1848, an important and influential Swedish chemist.

253

V. *Ueber künstliche Bildung des Harnstoffs;*
von F. Wöhler.

In einer früheren kleinen Notiz, die in dem III. Bande dieser Annalen abgedruckt ist, habe ich angegeben, dass beim Einwirken von Cyan auf flüssiges Ammoniak, ausser mehreren anderen Producten, auch Oxalsäure und eine krystallisirbare weisse Substanz entstehe, welche letztere bestimmt kein cyansaures Ammoniak sey, welche man aber dessen ungeachtet immer erhalte, so oft man auch versuche, z. B. durch sogenannte doppelte Zersetzung, Cyansäure mit Ammoniak zu verbinden. Der Umstand, dass bei der Vereinigung dieser Stoffe dieselben ihre Natur zu verändern schienen und dadurch ein neuer Körper entstände, lenkte von Neuem meine Aufmerksamkeit auf diesen Gegenstand, und diese Untersuchung hat das unerwartete Resultat gegeben, dass bei der Vereinigung von Cyansäure mit Ammoniak Harnstoff entsteht, eine auch in sofern merkwürdige That-sache, als sie ein Beispiel von der künstlichen Erzeugung eines organischen, und zwar sogenannten animalischen, Stoffes aus unorganischen Stoffen darbietet.

Ich habe schon früher angegeben, dass man die oben erwähnte krystallisirte, weisse Substanz am besten erhält, wenn man cyansaures Silberoxyd durch Salmiak-Auflö-sung, oder cyansaures Bleioxyd durch flüssiges Ammo-niak zersetzt. Auf die letztere Art habe ich mir die, zu dieser Untersuchung angewendete, nicht unbedeutende Menge davon bereitet. Ich bekam sie in farblosen, kla-ren, oft mehr als zollangen Krystallen angeschossen, die schmale rechtwinklige, vierseitige Säulen, ohne be-stimmte Zuspitzung, bildeten.

Mit kaustischem Kali oder mit Kalk entwickelte die-ser Körper keine Spur von Ammoniak, mit Säuren zeigte er durchaus nicht die so leicht eintretenden Zersetzungs-

254

Erscheinungen der cyansauren Salze, nämlich Entwicke-lung von Kohlensäure und Cyansäure, und eben so wenig fällte er, wie es ein wirkliches cyansaures Salz thut, die Blei- und Silbersalze: er konnte also weder Cyansäure noch Ammoniak als solche enthalten. Da ich fand, dass bei der letztgenannten Entstehungsart desselben kein an-deres Product mitgebildet wurde, so stellte ich mir vor, es könne bei der Vereinigung von Cyansäure mit Ammoniak eine organi-sche Substanz, und zunächst vielleicht ein den vegetabi-lischen Salzbasen ähnlicher Stoff entstehen; ich stellte daher aus diesem Gesichtspunkte einige Versuche über das Verhalten der Säuren zu dem krystallisirten Körper an. Er verhielt sich aber indifferent gegen dieselben, die Salpetersäure ausgenommen, welche in der concen-trirten Auflösung dieses Stoffes sogleich einen, aus glän-zenden Krystallschuppen bestehenden Niederschlag bil-dete. Diese Krystalle zeigten, nachdem sie durch mehr-maliges Umkrystallisiren gereinigt worden waren, sehr saure Charactere, und ich war schon geneigt, sie für eine eigenthümliche Säure zu halten, als ich fand, dass sie, bei der Neutralisation mit Basen, salpetersaure Salze gaben, von denen sich durch Alkohol der krystallisir-bare Stoff mit allen Charakteren, die er vor der Ein-wirkung der Salpetersäure hatte, wieder ausziehen liess. Diese Aehnlichkeit im Verhalten mit dem Harnstoff ver-anlasste mich, vergleichende Versuche mit vollkommen reinem, aus Urin abgeschiedenen Harnstoff anzustellen, aus denen ganz unzweideutig hervorging, dass Harnstoff und jener krystallisirte Körper oder das cyansaure Am-moniak, wenn man es so nennen könnte, vollkommen identische Stoffe sind.

Ich führe das Verhalten dieses künstlichen Harnstoffs nicht weiter an, da es vollkommen mit dem übereinkommt, wie es, nach den Angaben von Proust, Prout u. A., von dem Urin-Harnstoff in den Schriften zu finden ist.

1 *Annalen der Physik und Chemie* (Annals of Physics and Chemistry), 12, pp. 253–256 (1828).
2 *Annalen der Physik und Chemie*, 3, p. 177 (1825).
3 An adult human excretes about 30 g of urea in his urine per day.

116

And Berzelius answered on the 7th of March of the same year:

"...truly, you have mastered the art of finding the stepping stones to an immortal name. Aluminium and artificial urea; indeed, two very different substances following so closey upon one another! My dear Sir, they will be wound as jewels into your laurel wreath, and should the quantity of the artificial kind prove insufficient, one can easily supply the lack from any chamber pot ... It is an important and pleasing discovery that you have made, my dear Doctor, and I was indescribably delighted to hear of it...

Affectionate regards
Berzelius."

In the same year, Wöhler published his results in the *Annalen der Physik und Chemie* (Annals of Physics and Chemistry) under the title: "Concerning the Artificial Production of Urea"[1] (Ueber künstliche Bildung des Harnstoffs) in which he speaks of the unexpected result "that the combination of cyanic acid and ammonia produces urea, a remarkable effect in that it is an example of the artificial creation of an organic, even so-called animalistic, substance from inorganic compounds." To rephrase the matter in modern terms, Wöhler had obtained the organic compound urea, the diamide of carbonic acid, by evaporating an aqueous solution of ammonium cyanate (= an inorganic compound!):

$$NH_4^+ \, [\overline{|O} - C \equiv N|]^- \rightleftharpoons \begin{array}{c} H_2N \\ H_2N \end{array} \!\!\! > C = \overline{O}$$

Despite the fact that in 1825 Wöhler had reported at length his synthesis of oxalic acid[2], a common organic compound from inorganic compounds, it was the synthesis of urea, that most important of all metabolites[3], that caused a sensation in the scientific world. What had happened? Wöhler had synthesized a compound which, according to the scientific dogma of his day, only a living organism could possibly produce. Wöhler's success showed that there is no need for a special "life force" or *vis vitalis* in order to obtain organic compounds. Thus, this synthesis of urea contributed to the destruction of the idealistic philosophical tenets of vitalism, a theory that postulated the prerequisite existence of non-material forces which had to influence "dead" (inorganic) matter in order to turn it into "living" (organic) matter.

With these and other achievements, Wöhler contributed to the founding of a truly scientific organic chemistry and also became one of the founders of biochemistry, the discipline that concerns itself with the chemistry of life. One of the main branches of modern biochemistry is the study of the biological polymers, the proteins and the nucleic acids. Research into the chemistry of these vital polymers involves the study of their structure, of the synthetic process which produces them and of their functions in the organism. The biochemistry of these polymers is also known as molecular biology. On the following pages we will take a closer look at proteins and nucleic acids.

Proteins—the Biologically Multi-Functional Polymers

The term *protein* was coined by Berzelius in 1838 and is derived from the Greek word *proteuo* which means "I am the first". The name indicates the rank which 19th-century chemists assigned to the proteins among other chemical compounds. Today, chemists are inclined to grant equal status to the nucleic acids (see the section "Nucleic Acids—The Key Substance of Life) since the synthesis of the proteins is dependent on the information encoded in the nucleic acids. These two kinds of bio-polymers are the essential molecular foundations of all life.

The Biological Functions of Proteins

In all organisms, proteins perform a variety of biological functions. We will discuss a number of these functions as they occur in the human body, an organism which contains a very large number of different kinds of proteins. Proteins are essential for human nutrition. The human organism requires proteins as they occur in foodstuffs such as meat, milk and eggs, in order to synthesize its own, body-specific proteins. Muscles, sinews, connective tissues, skin, hair, nerves, blood vessels, and the very blood itself, all these are made up of cells containing more or less protein. The functions of these various parts of the human organism are largely dependent upon, or determined by, their proteins. The most important component of erythrocytes, the red blood cells, is a protein, namely haemoglobin. To be more exact, haemoglobin is a conjugated protein, a chromo-

In 1828 F. Wöhler published his results in the Annalen der Physik und Chemie *12, pp. 253–256: "Ueber die künstliche Bildung des Harnstoffs" (Concerning the artificial production of urea). This was of fundamental importance for the development of an organic synthesis chemistry.*

protein. A conjugated protein is a protein which also consists of other components such as metal ions, carbohydrates, dye-groups, or nucleic acids. Haemoglobin is capable of bonding molecular oxygen (O_2) from inhaled air, transporting it along the blood vessels, and releasing it at the appropriate oxygen-requiring sites in the organism. The human organism requires oxygen for the oxidation of nutrients, a process from which the body obtains its energy.

Other proteins, namely the *enzymes*, function as bio-catalysts and are responsible for catalyzing a vast number of chemical reactions in the body. Thus, proteins and fats ingested in food are first broken down into simpler compounds which are then used as the building blocks for the construction of the body's own proteins and lipids. Both the metabolic and the synthesis processes require proteins as catalysts for the necessary chemical reactions.

The protein insulin performs a most important function in the human organism in that it regulates the oxidation of glucose in the blood. A dificiency of this protein results in the well-known disease, diabetes mellitus, which, as is generally known, is treated with regular injections of insulin. Special proteins in the blood serum, known as serum proteins, protect the human organism against specific bacteria and viruses. Certain other serum proteins are responsible for blood coagulation and thus play an absolutely vital part in the healing process of wounds and lacerations. Proteins also transport hormones and metabolic products of all kinds. The effectiveness of many drugs and medicines is dependent upon transport by such protein molecules.

How is it possible for one type of substance to perform so many and such varied functions? The answer can only be found in the chemical constitution and the geometric structure of the proteins. The clarification of the total structure—the constitution and geometry—of proteins has only been achieved in the last few decades through the use of very sophisticated structural and analytical methods.

The Primary Structure of Proteins

Already by 1902 there were some intuitive ideas about the basic structure of the proteins when F. Hofmeister[1] and E. Fischer[2] hypothesized that proteins consist of amino-acid chains which are connected in an acid-amide sequence. Proteins are polymers of α-amino-carbon acids, which are commonly known as α-amino-acids and which have the following chemical structure:

1 Hofmeister, Franz, 1850–1922, biochemist.
2 Fischer, Emil, 1852–1919, German chemist. Nobel Prize for Chemistry in 1902.

where R represents a hydrogen atom or a group of atoms such as $-CH_3$. Such groups are called side chains of the α-amino-acids. The

groups $-\overline{N} \overset{\diagup H}{\diagdown H}$ and $-C \overset{\overline{O}}{\diagup\diagdown \underline{O}-H}$ are the primary amino group and

the carboxyl group, respectively. Both are characteristic of organic acids. The primary amino group can be found attached to the so-called α-carbon atom. The α-C-atom, immediately next to the carboxyl group, is almost invariably asymmetrical when R does not represent a hydrogen atom. In principle, the asymmetry makes it possible for proteins to be constructed from either L- or D- or from both L- and D-amino-acids. Oddly enough, however, all naturally occurring proteins contain only L-α-amino-acids.

Human proteins are constructed of this kind of amino-acid. It is quite easy to imagine a human being made up entirely of D-amino-acids. The D-organism would function biologically just as well and in the same manner as the L-organism.

Unfortunately, he could not survive in our world. Why? Because all vegetable and animal matter which he would have to ingest in order to fulfil his protein needs would contain only the L-α-amino-acids whereas he would require D-α-amino-acids. Why, then, do all natural proteins consist only of L-α-amino-acids? That is a question concerning the chemical evolution of amino-acids to which no satisfactory answer has as yet been found and which is still a topic of scientific dispute.

Proteins are formed through polycondensation of α-amino-acids. In principle, polymerization takes place in the following fashion:

$-H_2O \longrightarrow$

where R_1 and R_2 represent different side chains. The mechanism of this reaction is represented in a simplified form. When two amino-acids combine in this manner, a dipeptide (Greek: *di* = two) results.

The atomic grouping thus formed, $-\overset{\displaystyle H}{\underset{\displaystyle |O|}{\overset{|}{\underset{\|}{C}} - \underline{N}}} -$ is the so-called peptide

bond. As the drawing shows, such a dipeptide contains a carboxyl group and an amino group, just like the amino-acids. The reaction of these end groups with further amino-acids can result in tri-, tetra-, penta-, etc.-peptides (from the Greek: *tri* = three, *tetra* = four, *penta* = five). If more than ten amino-acids combine in this manner, we call the resulting chain-form polymers poly-amino-acids or polypeptides.

These polypeptide chains are the fundamental constituents of the proteins. The primary structure of a particular protein is characterized by the sequence and the number of amino-acid building blocks. In 1955, F. Sanger[1] et al. reported the first complete analysis of the primary structure of a natural protein. The protein in question was the pancreatic hormone, insulin. Up till now the primary structures of approximately 900 natural proteins have been established.

To date, there is no agreement as to the definition of proteins according to the number of their amino-acid building blocks. In general, we speak of proteins when the polypeptide chain contains more than 50 amino-acid units, i.e. when n is greater than 50. Proteins have also been defined as polypeptides with a molecular mass between 5,000 and several million.

Nature uses only 20 different amino-acids in its synthesis of proteins. These are known as proteinogenic α-amino-acids, and differ

120

1 Sanger, Frederick, born 1918, British chemist. Nobel Prize for Chemistry in 1958 and in 1980.

only in their R-groups. Table 4 gives the names and commonly used abbreviations for these amino-acids and also gives the structures of their side chains.

Table 4. Proteinogenic α-amino-acids

$$\begin{matrix} H \\ H \end{matrix} \!\!\! \rangle \overline{N} - \overset{\overset{\displaystyle H}{\displaystyle |}}{\underset{\displaystyle R}{C}} - C \!\!\! \langle \!\!\! \begin{matrix} \overline{O} \\ \overline{O} - H \end{matrix}$$

Name	Abbreviation	Structure of side chain R
Glycine	Gly	$-H$
Alanine	Ala	$-CH_3$
Serine	Ser	$-CH_2OH$
Threonine	Thr	$-CH(OH)-CH_3$
Cysteine	Cys	$-CH_2SH$
Valine	Val	$-CH(CH_3)_2$
Leucine	Leu	$-CH_2-CH(CH_3)_2$
Isoleucine	Ile	$-CH(CH_3)-CH_2-CH_3$
Methionine	Met	$-CH_2-CH_2-S-CH_3$
Phenylalanine	Phe	$-CH_2-\bigcirc$
Tyrosine	Tyr	$-CH_2-\bigcirc-OH$
Tryptophan	Trp	
Histidine	His	
Lysine	Lys	$-CH_2-CH_2-CH_2-CH_2-NH_2$
Arginine	Arg	$-CH_2-CH_2-CH_2-NH-C(NH)NH_2$
Glutamic acid	Glu	$-CH_2-CH_2-COOH$
Glutamine	Gln	$-CH_2-CH_2-CONH_2$
Aspartic acid	Asp	$-CH_2-COOH$
Asparagine	Asn	$-CH_2-CONH_2$
Proline	Pro	complete structure

For proline, the entire structure is given since it varies somewhat from the norm. In this amino-acid, the side chain proceeding from the α-C-atom links with the amino group to form a five-membered ring.

The number of different natural proteins in existence is today estimated to lie somewhere between 10 billion and 1 trillion. One reason for this vast number is the twenty different α-amino-acids which can be combined in any way. The other explanation of this huge number is the fact that the polypeptide chains can vary infinitely when it comes to length. Primitive organisms, such as bacteria, possess only a few thousand different proteins; the human organism, on the other hand, contains several million different kinds of proteins. In order to understand the structure of the proteins it is very important to comprehend the geometry of the peptide bond. The peptide bond is, in fact, one of the most thoroughly studied of the chemical bonds.

Bond lengths and bond angles in peptides.

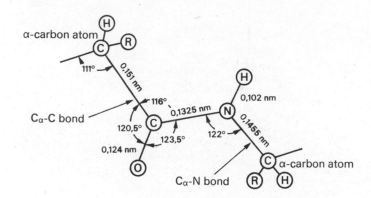

Secondary Structures of Proteins

The peptide bond shows a very interesting characteristic: it is a planar structure, i.e. all four atoms (O, C, N, H,) are located on a common plane. This means that rotations within the chain can take place only about the C_α–N bond and about the C_α–C bond. When these rotations take place, C = O-groups and NH-groups of different amino-acid residues can come close to one another within the same peptide chain and form hydrogen bonds of the following type: $>C = O \cdots H\overline{N}<$.

The formation of such bonds or links limits the movement of the polypeptide chains and constrains them to acquire definite spatial structures which have been designated as secondary structures. Hydrogen

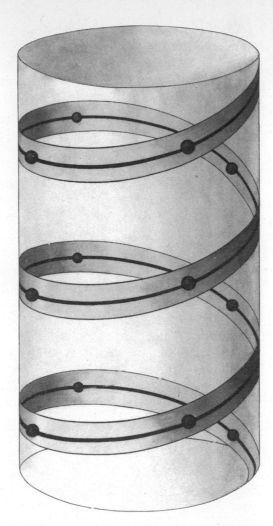

The helical structure (from R. E. Dickerson and I. Geis).

123

bonds of the above type can occur both within a single polypeptide molecule (intramolecular) and between two or more polypeptide molecules (intermolecular).

In the early 1950s, X-ray structural analyses of amino-acids, peptides, and polypeptides by Pauling and Corey led to the recognition that the polypeptide chains of certain proteins are organized in a spiral. This type of secondary structure has been given the name of *helix* (*helica:* Latin for "threaded"). This discovery was a decisive step in the elucidation of the spatial organization of proteins. The drawing above shows a helix which has four amino-acid residues (symbolized by dots) per one complete turn of the spiral.

The stability of these spirals is due to the intramolecular hydrogen bonds of the type $>C=O \cdots HN<$ between the turns of the helix.

The helices themselves are relatively rigid cylindrical structures in which the side chains R of the amino-acid component are located on the periphery. Helices differ in the number of amino-acids per turn. Pauling and Corey discovered a particularly stable and therefore frequent helical structure among the proteins, namely the α-helix. In this type of helix there are 3.6 amino-acids per turn. The α-helix is particularly stable because every peptide bond of the chains is also involved in the formation of a hydrogen bond. Like screws, helices can have either a right-hand or a left-hand thread. In all natural proteins with an α-helical secondary structure, however, the α-helices are invariably right-handed. This fact indicates that right-handed helices are naturally more stable than the left-handed type. The formation of stable α-helices depends upon the primary structure of the polypeptide chains. In other words, the type and sequence of the amino-acid building blocks of the chain decide whether or not α-helical structures can be formed.

Proteins of the keratin type are especially rich in α-helix structures. Keratins belong to the class of fibre proteins. They are the protein component of the skin, fur, hair, wool, claws, scales, nails, beaks, and horns of land vertebrates. Human hair keratin is made up of polypeptide chains with the spatial organization of α-helices. Three of these helices wind themselves into a triple helix, forming the next higher structure, the so-called proto-fibril.

Further "higher" structures which are involved in the make up of the actual hair are micro-fibrils, macro-fibrils and cells, but the structures of these components are not relevant here and will not be discussed.

Apart from hydrogen bonds, there are other ways of stabilizing or fixing of certain three-dimensional protein conformations. One of these is the formation of a disulphide bridge which depends on the presence of cysteine (see Table 4: Proteinogenic α-amino-acids, p. 121) in the polypeptide chain. The reactive mercapto groups (−SH) of the two cysteine building blocks can form a link if they split off the two hydrogen atoms. This link or bond will consist of two sulphur atoms; hence the name, disulphide bridge.

$$-SH + HS- \xrightarrow{-2H} \quad -S-S-$$

mercapto groups   disulphide bridge

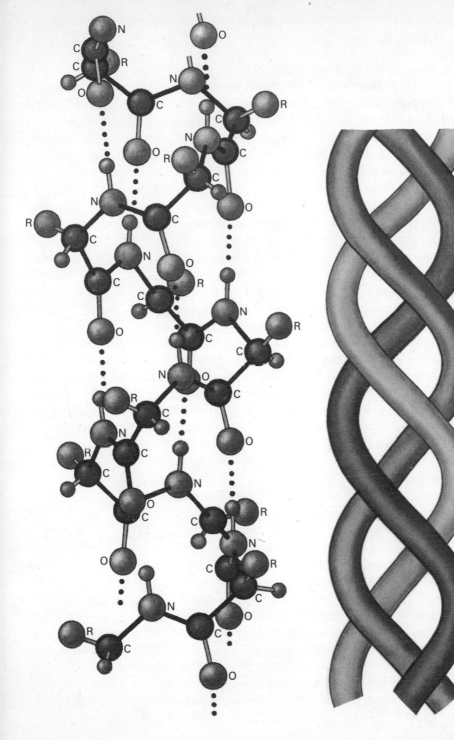

A spatial model of the α-helix (from A. Lehninger).

Proto-fibril

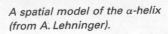

Such disulphide bridges can bond certain segments within a polypeptide chain (intrachain disulphide bridges) as well as two or more polypeptide chains (interchain disulphide bridges). Insulin is a protein containing both the above kinds of disulphide bridge.

Insulin consists of two polypeptide chains, the A chain, consisting of 21 amino-acids, and the B chain, consisting of 30 amino-acids. The two chains are linked by two interchain disulphide bridges. The intrachain disulphide bridge connects positions 6 and 11 in the A chain. An analysis of the insulin of different species of animals showed certain variations in the amino-acid sequence. However, these differences are limited to positions 8, 9, and 10 in the A chain and position 30 in the B chain. Thus, the insulins of cattle, sheep, and horses are only negligibly different. What is more, the insulin of dogs, pigs and sperm whales is identical.

Interchain disulphide bridges also occur with hair keratin. Here, neighbouring helices are bonded to one another with disulphide bridges; the internal structure of the helices themselves is maintained by way of intramolecular hydrogen bonds. The elasticity of the hair and wool keratins can be traced back to these basic structural characteristics. When stretched, the weak hydrogen bonds within the helices break. The result is that chains become broken and the material gives. The disulphide bridges almost 10 times stronger between the helices remain intact and resist the stretching. When the force that is stretching the polypeptide chain ceases, the intact disulphide bridges provide the main impulse for a return to the initial state in which the hydrogen bonds re-form. It is possible to break the disulphide bridges by reducing them to mercapto groups. Then the "free" polypeptide helices can easily be shaped, a process which takes place every time someone acquires a permanent wave. The shape of the curls is then fixed through oxidation, during which process disulphide bridges are re-formed in the newly styled hair.

A further important secondary structure of the fibre proteins is the pleated structure, also discovered by Pauling and Corey. Polypeptide chains will fold into parallel or antiparallel pleats when hydrogen bonds form *between* two or more polypeptide chains. An antiparallel pleated structure is characteristic of the protein fibroin, for instance. Fibroin is the most important protein component of silk.

Apart from fibre proteins, the globular proteins constitute one of the main classes of proteins. Nearly all enzymes are proteins of this type, as are many of the proteins with a transport function such as myoglobin and haemoglobin. The name is descriptive; "globular"

A-chain		B-chain
NH₂		NH₂
Gly 1		Phe 1
Ile		Val
Val		Asn
Glu		Gln
Gln		His
Cys 6		Leu
Cys — S — S —		Cys 7
Thr 8		Gly
Ser 9		Ser
Ile 10		His
Cys 11		Leu
Ser		Val
Leu		Glu
Tyr		Ala
Gln		Leu
Leu		Tyr
Glu		Leu
Asn		Val
Tyr — S —		Cys 19
Cys — S —		Gly
Asn 21		Glu
COOH		Arg
		Gly
		Phe
		Phe
		Tyr
		Thr
		Pro
		Lys
		Ala 30
		COOH

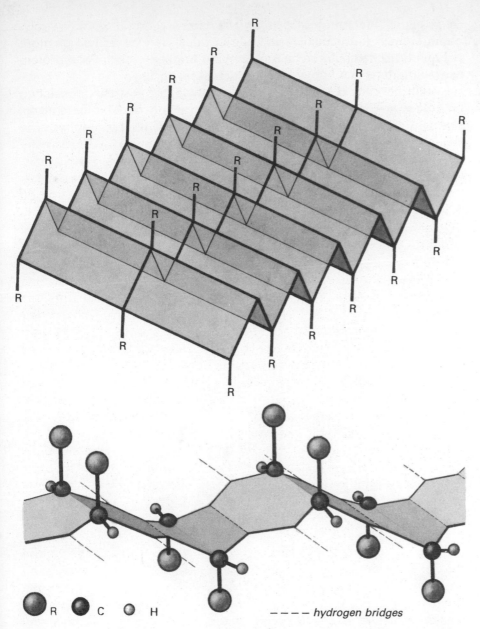

A pleated structure with three parallel chains (from A. Lehninger).

An antiparallel pleated structure (from R. E. Dickerson and I. Geis).

R C H — — — — *hydrogen bridges*

comes from the Latin (*globulus* = little ball) and refers to the fact that the polypeptide chains of these proteins are folded and wound to produce a ball-like shape.

Tertiary and Quaternary Structures of Proteins

In the 1950s, a team of researchers headed by J. C. Kendrew[1] in Great Britain achieved the first elucidation of myoglobin, a three-dimensional structure of a globular protein. This was made possible through the use of the then most modern methods of X-ray diffraction and the aid of computers in evaluating the sheer quantity data acquired. The structural model that was proposed in 1957 was then modified in the light of additional information. By the early 1960s, the exact spatial location of every atom in this immensely complicated

128 *The haeme group which occurs in both myoglobin and haemoglobin.*

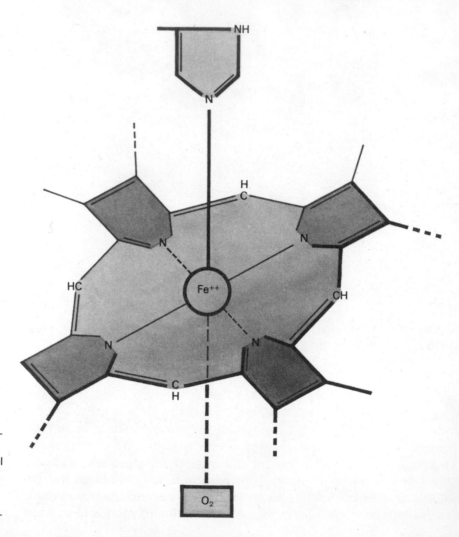

1 Kendrew, John C., born 1917, British biochemist. Nobel Prize for Chemistry in 1962, together with Max Ferdinand Perutz.

molecule had been established. An important and complementary contribution to this complete structural analysis was made by the parallel chemical analysis of the amino-acid sequence in this protein, which was carried out by A. B. Edmundson in the U.S.A.

Myoglobin is a relatively small protein. It consists of only a single polypeptide chain with 153 amino-acid building blocks. It also incorporates a dye-group, the haeme-group, so that myoglobin is, in fact, a composite protein, a proteid. Myoglobin is found in the muscles of vertebrates where it functions as an oxygen storing protein; it is also responsible for the red colour of the muscle tissue. The heart muscle of diving ocean mammals (whales and seals) is especially rich in myoglobin. Like haemoglobin, myoglobin can bind oxygen and release

130

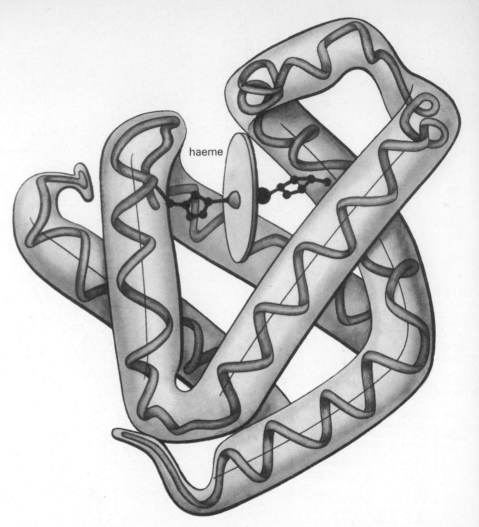

haeme

it when needed. The biological function of the two proteins is closely
linked to the structure of the haeme-group, the latter being that part
of the molecule which can reversibly bind an oxygen molecule. The
haeme-group consists of a ring system, protoporphyrin, which contains
an iron (II)-ion at its centre. This metal ion is completely saturated
with four heterocyclic rings from the protoporphyrin ring system
while at the same time being co-ordinated with a histidine residue
of the polypeptide chain and an oxygen molecule. 121 amino-acids
of the polypeptide chain are organized into eight, relatively straight,
α-helical segments which, in turn, are connected by crooked, non-
helical, and less orderly segments. With a bit of imagination, the

myoglobin molecule can be seen as a flattened ball. The total structure gives an impression of great irregularity, but it is nonetheless a defined and limited structure, the product of all the mutual interactions between the atoms which make it up. It is this rather messy looking structure which enables myoglobin to fulfil its biological function. This structural description which, on the one hand, defines the geometry of the secondary structural segments and, on the other, gives exact details as to the spatial organization of all the individual atoms in the molecule, is known as the tertiary structure of the proteins.

Related to myoglobin, but chemically and biologically much more complicated is haemoglobin, the red colouring substance in the blood. This protein is found in the erythrocytes (red blood cells) of vertebrates. Its biological function is to transport oxygen molecules from the lungs to the tissues. 100 ml of human blood, for example, contains 15 to 16 g haemoglobin which can bond approximately 21 ml oxygen (O_2). To be exact, the red colouring agent of adult human blood is made up of two types of haemoglobin: 96.5 to 98.5 % is the so-called haemoglobin A_1 and 1.5 to 3.5 % is haemoglobin A_2. The structural differences between these two types of haemoglobin are minor. Apart from two further types of haemoglobin, which occur normally in the human being during the embryonic phase, about 150 abnormal haemoglobin types have been found in the human organism. These abnormal haemoglobins usually differ from the normal protein only in minor deviations in the amino-acid sequence of the protein's polypeptide chain. In our examination of the structure of haemoglobin, we will concentrate on the structure of haemoglobin A_1.

Just like myoglobin, haemoglobin A_1 is actually a proteid. Its protein part consists of four polypeptide chains which are not interconnected by covalent bonds. Two of these polypeptide chains are made up of 141 amino-acids each and are known as α-chains. The other two polypeptide chains consist of 146 amino-acids each and are known as the β-chains. The primary structure of these four chains with their total of 574 amino-acids was first determined by Braunitzer. As to its secondary structure, about 70 % of the haemoglobin chains are organized into α-helices. The tertiary structure of the polypeptide chains of haemoglobin is very similar to that of the myoglobin chains. Even the haeme-group of the myoglobin is the same as that of the haemoglobin. Each of the four polypeptide chains is bonded to a haeme-group with a non-covalent bond. A total of four O_2 molecules can be reversibly bonded to a single haemoglobin molecule[1], one per chain, thus forming oxyhaemoglobin.

131

1 The term molecule is here used in the extended sense of species which are in fact constituted of several molecules.

The fact that haemoglobin consists of several and not only *one* polypeptide chain is of special interest. Such proteins, known as oligomeric proteins, characteristically organize their polypeptide chains, which are already fixed in their primary, secondary, and tertiary structures, into a further, spatially defined, arrangement known as the quaternary structure. Quaternary structural bonds are also non-covalent. Instead, weaker forms of bonding, such as hydrophobic bonding between non-polar side chains of the amino-acids, become effective. With haemoglobin, the four chains aggregate in an approximately tetrahedral configuration, forming a compact spherical particle with the dimensions $6.4 \times 5.5 \times 5.0$ nm. The four haeme-groups are located in separate "pockets" on the surface of the molecule. The

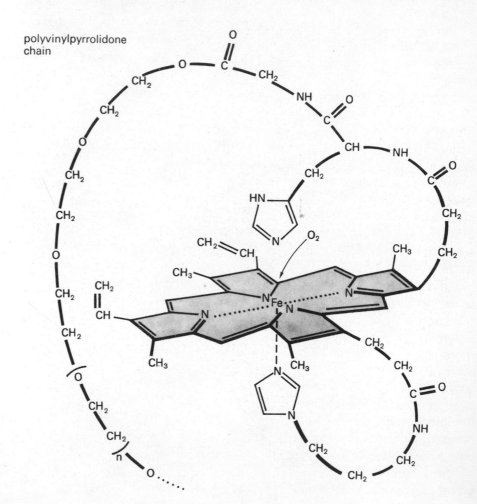

polyvinylpyrrolidone chain

A synthetic haeme polymer (from E. Bayer and G. Holzbach).

spatial structures of the monomeric and also of the oligomeric proteins consisting of only one polypeptide chain are no rigid structures but have some flexibility.

As has become evident in the example of haemoglobin, oligomeric proteins are aggregates of sub-units. In haemoglobin, for instance, the four polypeptide chains could be regarded as sub-units. There are, however, other ways of sorting out sub-units; one could create two sub-units, consisting of an α-chain and a β-chain each. Under certain environmental conditions, haemoglobin will even dissociate into two sub-units of two α-chains and two β-chains. Such sub-units and their formation are frequently associated with specific biological functions, such as when an oligomeric protein becomes effective as an enzyme.

With its maximum of four sub-units and a relative molecular weight of 64,500 haemoglobin is a midget among the oligomeric proteins. Let us take, in contrast, the protein segment of the tobacco mosaic virus. This oligomeric protein possesses a relative molecular weight of 38,000,000 and is constructed of 2,200 identical polypeptide chains. The elucidation of the tertiary and quaternary structures of the haemoglobin of humans and that of horses was achieved with the use of X-ray diffraction studies by M. F. Perutz[1] et al. in Great Britain. Their research was carried out at the same time and parallel to J. C. Kendrew's myoglobin structural analysis but took longer to complete because of the more complicated structure of the haemoglobin. Both structural analyses were carried out at the famous Cavendish Laboratory of Cambridge University, the same laboratory in which Watson and Crick[2] worked out the three-dimensional structure of deoxyribonucleic acid in 1953.

It has recently become possible to synthesize artificial haemepolymers[3] whose properties are very like those of the natural oxygen transporting proteins, myoglobin and haemoglobin. These artificial polymers are not polypeptides but the more easily synthesized polymers such as polyvinylpyrrolidones, polyethylene-oxide-bis-glycine ester (also known as polyethylene-glycol-bis-glycine ester), and certain polyurethanes. The main prerequisite for the functioning of such synthetic blood substitutes is the far-reaching structural identity of the functional centres (of the areas where the oxygen molecule is bonded) of the polymers with those of the myoglobin or haemoglobin. Another condition has good solubility in water. The importance of the synthesis of such polymers for the development of artificial blood plasmas is easy to see.

1 Perutz, Max Ferdinand, born 1914, Austrian chemist, resident in Great Britain since 1936. Nobel Prize for Chemistry in 1962 together with J. C. Kendrew.
2 See also the Section "Nucleic Acids—the Key Substance of Living Matter", Chapter IV.
3 E. Bayer and G. Holzbach, *Angewandte Chemie* (Applied Chemistry), 89, (1977) No. 2, pp. 120–122.

Proteins as Bio-Catalysts: The Enzymes

We have already mentioned the important role of proteins as bio-catalysts. Proteins with this function are called enzymes. They are essential to all the chemical reactions within the entire organism, i.e. to the entire metabolic process. Enzymes control both the speed and the direction of all chemical reactions within biological systems. There is an enzyme for every specific function. That enzymes are proteins was first established in 1926. Today, we know more than 2,000 different enzymes, only approximately 200 of which have so far been isolated in a pure, usually crystalline form. All metabolic procedures such as the oxidation of lipids, the synthesis or decomposition of proteins and nucleic acids, and all fermentation processes are dependent upon the action of enzymes.

General progress in the elucidation of the spatial conformation of proteins has also led to a better understanding of the relationship between structure and function in the enzymes. As far as the molecular working mechanism of the enzymes is concerned, Emil Fischer already proposed the following model. According to Fischer, the active centre of an enzyme has a spatially rigid shape and can thus only influence those molecules which fit into this active centre like a key into a lock. In other words, they can only affect those molecules which have a complementary shape. The molecules which attach themselves to the enzyme molecules are called the substrate molecules. Together with the enzyme molecules, they form the enzyme-substrate complex, which gives the reaction product and the enzyme after a certain, usually brief, reaction time. A molecule of the enzyme catalase, for instance, can catalyze the decomposition of five million of its substrate molecules in one minute, a tremendous reaction rate.

Another theory of the mechanism of enzyme action is the "induced fit" theory of Koshland. This theory postulates that initially the substrate molecule induces certain structural modifications in the enzyme molecule. The enzyme molecule is transformed into a more active high energy state, and only after this has occurred does the formation of the enzyme-substrate complex take place. While this enzyme-

The key-and-lock theory

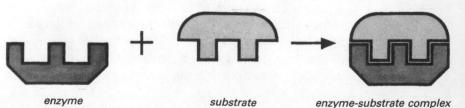

enzyme   substrate   enzyme-substrate complex

substrate complex is in existence, the enzyme molecule so modifies the structure of the substrate that it proceeds from a relatively stable state to a less stable and more reactive state.

Enzymes are classified into a number of categories according to the basic chemical function which they control. Thus, we have oxidoreductases, hydrolases, transferases, lyases, isomerases, ligases and other enzyme categories. Oxidoreductase enzymes are functional in oxidation-reduction processes; they control the transfer of hydrogen and electrons. The hydrolase enzymes catalyze the hydrolysis of proteins and other compounds. Fig. 44 shows the molecular model for bovine trypsin, an enzyme of the hydrolase category.

Since enzymes are chemically so very efficient, it is easy to see why a great deal of work is being done in order to enable industry to use them in a controlled manner in a large number of areas. Mankind has long exploited enzymatic processes in the production of things like beer, bread or cheese. All fermentation processes involving moulds, yeasts and bacteria are really using these micro-organisms by exploiting their enzymes in order to make their product (see also Chapter V). The term "fermentation agent" is actually synonymous with the term "enzyme"; both terms originally referred to the process which produces sour dough. Enzymes are finding more and more uses in all the branches of industry such as the chemical and pharmaceutical industries. They are also used in the manufacture of detergents, textiles, paper and in the processing of foodstuffs. A particularly good example is the use of enzymes for the economically rational production of complicated pharmaceuticals and organic substances. They are also used to tenderize meat and preserve perishable foodstuffs and for many other things. A major problem in the exploitation of enzyme functions is the enzyme's perishability. Enzymes are, after all, proteins and thus very sensitive to temperature changes. One method to increase the stability of natural enzymes is to join them to "carriers", such as cellulose or synthetic macromolecules. Chemists are currently working on this problem, trying to produce artificial enzymes with properties that do not have the weaknesses of natural enzymes.

The Chemical Synthesis of Proteins

At the beginning of this century, Emil Fischer had already managed to synthesize artificial peptides. To be exact, he succeeded in constructing polypeptide chains up to a length of 16 amino-acids. Such syntheses were exceedingly costly. Only recently have more economical methods for such syntheses been developed. R. B. Merrifield and others have been mainly responsible for this advance. For peptide syntheses, the main problem consists of creating a reaction environment in which only the functional groups required for the formation of the desired peptide bond will react with one another. The problem becomes clear if we consider, as an example, the many possibilities of peptide bonding between the two following α-amino-acids:

$$\underset{\text{amino-acid 1}}{HOOC - \underset{\underset{R_1}{|}}{\overset{\overset{H}{|}}{C}} - NH_2} + \underset{\text{amino-acid 2}}{HOOC - \underset{\underset{R_2}{|}}{\overset{\overset{H}{|}}{C}} - NH_2}$$

There exist the following reaction possibilities:

1. The amino groups of amino-acid 1 will react with the carboxyl groups of amino-acid 2, forming dipeptide 1.
2. The amino groups of amino-acid 2 react with the carboxyl groups of amino-acid 1, forming dipeptide 2.
3. The amino groups of amino-acid 1 react intramolecularly with the carboxyl groups of amino-acid 1, forming dipeptide 3.
4. The amino groups of amino-acid 2 react intramolecularly with the carboxyl groups of amino-acid 2, forming dipeptide 4.

The reaction mixture will, therefore, contain four different dipeptides. Moreover, certain functional groups such as $-SH$ or $-OH$, which occur in the amino-acid side chains might enter into the reaction process and further increase the number of reaction products. To isolate particular dipeptides from such a reaction mixture involves a high technical expenditure. The actual amount of the desired dipeptide is proportionally reduced through the existence of the competing reaction products in the original mixture. These synthesis problems can be overcome if all the functional groups which we do not want to enter into a peptide bond can be stopped from doing so. The method for achieving this is to provide such functional groups with a "guardian", i.e. to bond them to an inhibitor which prevents them from reacting (see p. 137).

After the desired peptide bond has been formed, these blocking groups can be split off again. The repeated use of this process allows us to construct polypeptides systematically. In the process developed by Merrifield, the carboxyl group of the first amino-acid is bonded to a particle of an insoluble resin, the amino group remaining free. In the second amino-acid necessary for the peptide bond, the amino group is bonded to an inhibitor and the carboxyl group of amino-acid 2 is left free so that it can form a peptide bond with the amino group of the first amino-acid. The reaction product is a dipeptide with an attached resin particle and a blocked amino group. Once the blocking group has been removed, the peptide bonding process can be repeated. It is a considerable practical advantage that the resin particles are insoluble and can be filtered off in the liquid reaction stage. Reaction components are then separated from the resin particles with appropriate solvents and the resin particles re-used in the further synthesis process. Only when the polypeptide chain has achieved the desired length and the particular synthesis is complete is the resin finally removed. Compared to older methods, this new polypeptide synthesizing technique not only saves a great deal of time, it can also be fully automated. Thus, it has become possible to synthesize the A-chain of insulin with its 21 amino-acid residues in 8 days, and the B-chain with its 30 amino-acid residues within 11 days. A ribonuclease of 124 amino-acids, the natural product of a bovine pancreas, has also been artificially synthesized using this elegant method. Another method for constructing polypeptides is by way of fragment condensation, a process in which smaller peptide units are joined together to form larger ones. The most recent developments involve a combination of elements of the Merrifield method and of the fragment condensation method in the so-called liquid-phase method. This technique synthesizes polypeptides in liquid phase and has led to further advances in the chemical synthesis of polypeptides.

Nucleic Acids—the Key Substance of Living Matter

In 1869, Friedrich Miescher (1844–1895), a Swiss working in Tübingen, used a soda solution to extract a phosphorous-rich substance from the nuclei of human pus cells. He called this substance "nuclein" derived from the Latin word *nucleus*. "After experiments with other tissues," he came to the conclusion that "an entire family of slightly differing phosphorous containing substances will be found, which, as the group of nuclein compounds, will probably be comparable with the family of proteins."

In 1874, Miescher reported that he had split the nuclein from the sperm of Rhine salmon into a protein molecule and a non-protein molecule which he identified as being a multifunctional acid. In 1889, this acid was renamed nucleic acid by another scientist. Miescher had had to abandon his research into nucleic acid because of his many duties as professor of physiology in Basel. He was only able to return to his interrupted work in 1887. Subsequently, he tried to solve two problems concerning nucleic acid: firstly to analyze the substance in order to establish its exact chemical formula and secondly, to solve the problem of the bio-synthesis of nucleic acid. Given the complexities of this compound, the solutions to both problems were beyond the experimental capabilities of his day and Miescher failed

138

The five most important nitrogenous bases which occur in nucleotides. (In these and all following formulae free electron pairs are not represented.)

2,4-dioxopyrimidine
uracil (U)

2-oxo-4-aminopyrimidine
cytosine (C)

5-methyl-2,4-dioxopyrimidine
thymine (T)

2-amino-6-oxopurine
guanine (G)

6-aminopurine
adenine (A)

in his endeavours. A valuable contribution to the structural analysis of nucleic acid was made in the 19th century by A. Kossel in Berlin. He and his students determined the structure of both the purine bases, adenine and guanine, and of the pyrimidine bases, uracil, cytosine, and thymine, which they had extracted from nucleic acid by means of hydrolysis.

The Chemical Structure of Nucleic Acids

In the first half of the 20th century, our knowledge concerning the chemistry of nucleic acids advanced step by step. Chemists learned to distinguish between ribonucleic acids and deoxyribonucleic acids. It became clear that the nucleic acids are composite molecules made up of the so-called mononucleotides. It was further recognized that these nucleotides consist of a nitrogenous base, a C_5-sugar (a pentose) and a phosphoric acid residue.

D-ribose   2-deoxy-D-ribose

The C_5-sugars (pentoses) occurring in the nucleotides.

In our study of the substance we will proceed systematically and construct the nucleic acids step by step without considering the actual biosynthesis process for the compound. We will begin with the nitrogenous bases, the above named pyrimidine bases uracil, cytosine,

Adenosine—a ribonucleoside.

Three hydroxyl groups of the sugar component in positions 2′, 3′ and 5′ can be esterified by phosphoric acid.

2′-deoxyadenosine, a 2′-deoxy-ribonucleoside.

Two hydroxyl groups of the sugar component in positions 3′ and 5′ can be esterified by phosphoric acid.

and thymine, and the purine bases adenine and guanine, all of which are commonly designated by the letters U, C, T, A, and G. If we join a pyrimidine base through position 1 on the ring, or a purine base by position 9 to a pentose, then the product is not a mononucleotide, but it is a precursor, a nucleoside. Two types of nucleosides are important for our purposes, the ribonucleosides and the 2'-deoxyribonucleosides. The ribonucleosides contain a D-ribose as their pentose; the 2'-deoxyribonucleosides have the 2-deoxy-D-ribose instead. The latter, in contrast to the D-ribose, has no hydroxyl group on its carbon atom 2. The ribonucleosides, which are derived from the bases adenine, uracil, cytosine, thymine and guanine, are popularly known as adenosine, uridine, cytidine, thymidine, and guanosine. For the respective 2'-deoxyribonucleosides, the term 2'-deoxy- is prefixed to the appropriate term, e.g. 2'-deoxy-adenosine.

There is only one more step from the nucleosides to the mononucleotides. The nucleosides must be joined to a phosphoric acid. This is accomplished through the reaction of the pentose hydroxyl group with phosphoric acid. When alcoholic hydroxyl groups—and such are the groups found in pentose—react with acids, esters are usually formed. Mononucleotides are, therefore, phosphoric acid esters of nucleosides, or, in other words, nucleoside phosphates. Naturally occurring nucleotides can exist as mono-, di- and tri-phosphates. Mononucleotides, whose sugar component is a 2-deoxy-D-ribose, are known as deoxyribonucleotides. Analogously, there are the ribonucleotides with the D-ribose. Since ribonucleosides have three hydroxyl groups and deoxyribonucleosides only two hydroxyl groups which can be involved in esterification, the phosphate group of the nucleotides can be bound to any one of these hydroxyl groups. This results in the formation of various nucleotide types: 3'- and 5'-deoxyribonucleotides and 2'-, 3'- and 5'-ribonucleotides. In cells, 5'-nucleo-

$$HO-\overset{\overset{\displaystyle OH}{|}}{\underset{\underset{\displaystyle O}{\|}}{P}}-OH \qquad \textit{monophosphoric acid}$$

$$HO-\overset{\overset{\displaystyle OH}{|}}{\underset{\underset{\displaystyle O}{\|}}{P}}-O-\overset{\overset{\displaystyle OH}{|}}{\underset{\underset{\displaystyle O}{\|}}{P}}-OH \qquad \textit{diphosphoric acid}$$

$$HO-\overset{\overset{\displaystyle OH}{|}}{\underset{\underset{\displaystyle O}{\|}}{P}}-O-\overset{\overset{\displaystyle OH}{|}}{\underset{\underset{\displaystyle O}{\|}}{P}}-O-\overset{\overset{\displaystyle OH}{|}}{\underset{\underset{\displaystyle O}{\|}}{P}}-OH \qquad \textit{triphosphoric acid}$$

140

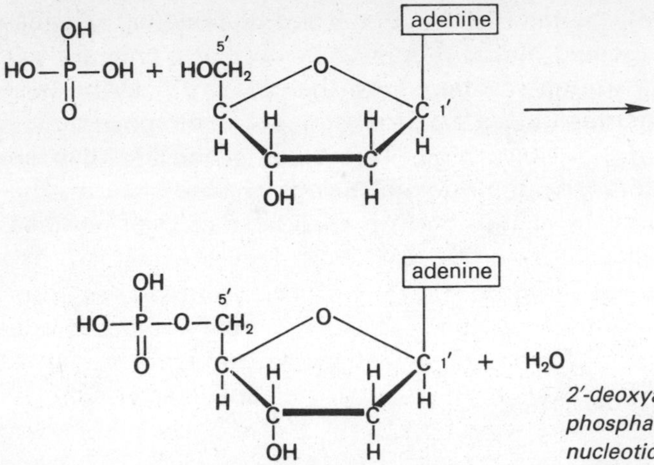

2'-deoxyadenosine-5'-mono-
phosphate—a deoxyribo-
nucleotide

tides are especially prevalent. Among ribonucleotides, these are the 5'-phosphates of adenosine, guanosine, cytidine, and uridine. The especially frequent 5'-phosphates of the deoxyribonucleotides are deoxyadenosine, deoxyguanosine, deoxycytidine, and deoxythymidine. For these compounds, abbreviations have been introduced. For example, adenosine-5'-monophosphate is shortened to AMP. If the compound is an ester of diphosphoric acid, an adenosine-5'-diphosphate, the abbreviation is ADP; similarly, a triphosphate becomes ATP. For the deoxyribonucleotides, these abbreviations are prefixed with a "d" in order to differentiate them from the ribonucleotides; e.g. dAMP stands for deoxy-adenosine-monophosphate. Depending upon the base involved (A, G, C, U, T), the pentose (a ribose or a deoxyribose) and phosphoric acid (mono, di-, or tri-phosphoric acid) the important 5'-nucleotides are identified by the following abbreviations:

5'-ribonucleotides (ribonucleoside-5'-mono-, di-, and triphosphates)			5'-deoxyribonucleotides (deoxyribonucleoside-5'-mono-, di-, and tri-phosphate)		
AMP	ADP	ATP	dAMP	dADP	dATP
GMP	GDP	GTP	dGMP	dGDP	dGTP
CMP	CDP	CTP	dCMP	dCDP	dCTP
UMP	UDP	UTP	dTMP	dTDP	dTTP

When many mononucleotides are joined to one another, polynucleotides are formed, also known as nucleic acids. Polynucleotides which are constructed of ribonucleotides are known as ribonucleic acids or RNA. The deoxyribonucleic acids (DNA) are made up of deoxyribonucleotides. The bonding together of the mononucleotides into the poly-nucleotides (= the nucleic acids) takes place by way of a phospho-di-ester link from the 3'-position of the one nucleotide to the 5'-position of the other. This bonding site of the nucleotides is made possible by the fact that the phosphoric acid in position 5' still possesses reactive protons (H^+) and the 3'-position in both the ribonucleotides and the deoxyribonucleotides is occupied by a hydroxyl group which allows further esterification.

Polynucleotides = nucleic acids

Deoxyribonucleic acid (DNA)— a single-stranded molecule

142 *Ribonucleic acid (RNA) (Illustration taken from Biochemie by A. L. Lehninger, Weinheim, 1975.)*

phospho-diester link ▶

If we look closely at the nucleic acids, DNA and RNA, it becomes evident that the basic nucleotide chain of these biopolymers consists of alternating pentose and phosphoric acid residues. The nitrogenous bases are joined to the off-side of the pentoses:

The structural model reproduced reflects our chemical knowledge concerning nucleic acids up to the early 1950s.

The Biological Function of Deoxyribonucleic Acid

The biological function of nucleic acids remained a mystery until the mid-1940s. DNA had been recognized as the constituent material of the chromosomes[1], but the general scientific opinion of the day regarded DNA as being biologically uninteresting, not to say insignificant. A change in this attitude did not come about until 1944 when the American bacteriologist O. T. Avery published his remarkable papers. Avery's starting point had been the work of the British bacteriologist Griffith, who had made an astonishing discovery in his study with pneumococci bacteria in 1927. It so happens that some strains of pneumococci form a "smooth" surface as they expand into colonies, and these are pathogenic. This is in direct contrast to other pneumococci strains whose colonies form "rough" surfaces and which do not cause disease. Careful examination showed that the "smooth" pathogenic pneumococci are encapsulated in a membrane consisting of polysaccharides. The "rough" innocuous bacteria do not have such a membrane. Because of their membrane, the pathogenic bacteria obviously resist detection by the host's immune system and can spread in an organism and so can become pathogenically effective. Griffith's experiments showed that the ability to encapsulate could be transferred from smooth bacteria killed by exposing them to heat to living "rough" pneumococci. Griffith carried out this transferral of the genetic ability to encapsulate through the medium of living mice. When mice were injected simultaneously with living "rough" bacteria (harmless) and dead "smooth" bacteria (equally harmless, since the bacteria were dead), the mice proceeded to develop living "smooth" bacteria which proceeded to kill the mice. At a later date, it became possible to show that a direct mixing of dead "smooth" bacteria and living "rough" ones in a certain culture medium would result in growing colonies of living "smooth" bacteria. The experiment was even carried a step further. In order to exclude the possibility that the dead "smooth" bacteria were simply not dead

1 According to the classical definition, chromosomes are light-microscopically visible morphological structures in the nuclei of eucaryotic organisms. The modern definition stresses the chromosomal function of carrying the genetic information in both procaryotic organisms and in the nuclei of eucaryotes. Cells of eucaryotic organisms are primarily characterized by their possession of a nucleus. Thus, unicellular organisms such as flagellates and green algae are eucaryotes as are all higher forms of plant and animal life as well as man. Cells of procaryotic organisms, such as blue algae and bacteria, are morphologically of a simpler structure. Unlike in eucaryotes, these cells do not have a nucleus separated from the rest of the cytoplasm by a membrane. Instead, procaryotes have the so-called nucleoides, precursors of a proper cell nucleus, which are DNA bonded to proteins, and which fulfil the same genetic function.

enough or could somehow be re-animated by their harmless "rough" colleagues, an extract of the dead bacteria was prepared which actually contained no intact dead bacteria. The ability to encapsulate was also transferred by this extract to living "rough" bacteria; moreover, the acquired genetic ability was retained by the formerly "rough", now "smooth", bacteria in successive generations. Avery made it into the goal of his work to isolate the substance which was responsible for transferring the genetic trait. He and his co-workers began with a cell-free extract of dead "smooth" bacteria which they prepared and purified in such a manner that its ability to transform the recipients of living "rough" bacteria remained unimpaired. The efforts of Avery and his team of co-workers resulted in a basic discovery for molecular biology: the substance responsible for transforming the genetic trait of the receiving bacteria is deoxyribonucleic acid. It is this substance which transfers the genetic information. It must, therefore, be the material through which the genetic information of bacteria is passed on. Up to the time of Avery's discovery, scientific opinion had considered that the genetic material of living organisms must be some form of protein. That chromosomes do play a part had been recognized, but the chromosomes had been considered as mere carriers of the genetic material. After Avery's research, it became clear that all genes are constituted of DNA since the chromosomes are made up of DNA.

The tendency of geneticists to experiment with microorganisms such as bacteria or viruses when they are doing genetic research has several reasons. For one thing, bacteria are monocellular creatures which, in comparison to higher organisms, have a fairly simple construction and are easier to research. For another, bacteria multiply rapidly and are very small. This means that it is possible to carry out experiments involving a very large number of organisms in very limited space. That the experiments should involve a very large number of organisms is essential in order to arrive at conclusions that are statistically significant. In other words, the large numbers are a safeguard against individual variations becoming decisive to an experimental result.

The concept of viruses as a kind of "naked gene" has led to intensive and extensive research with these particles which can reproduce themselves, but only in a living host cell. Viruses occur as cell parasites in all kinds of organisms, such as plants, animals and even microorganisms like bacteria. Experiments carried out in 1952 by Al Hershey and Martha Chase with bacterial viruses (bacteriophage) produced further proof that DNA is the basic genetic material.

When bacteria reproduce, a single cell undergoes cell division, producing two identical daughter cells. During this process, the genes must duplicate themselves. If we assume that genes are made of DNA, then the DNA matter in the cell must double at some point before cell division. Exactly how does the DNA duplicate itself? This was the question which posed itself to the young American biologist James D. Watson in the early 1950s. The then known facts concerning the structure of the DNA gave him no answer.

"Nonetheless, in contrast to the proteins, the solid chemical facts known about DNA were meager. Only a few chemists worked with it and, except for the fact that nucleic acids were very large molecules built up from smaller building blocks, the nucleotides, there was almost nothing chemical that the geneticist could grasp at."[1]

"That was not to say that the geneticist themselves provided any intellectual help. You would have thought that with all their talk about genes they should worry about what they were. Yet almost none of them seemed to take seriously the evidence that genes were made of DNA. This fact was unnecessarily chemical."

"On the other hand, it was equally obvious that I had not done anything which was going to tell us what a gene was or how it reproduced. And unless I became a chemist, I could not see how I would."

The Elucidation of the Three-Dimensional Structure of DNA

James D. Watson, who studied ornithology in his undergraduate years and spent some time in post-graduate phage research went to Europe on a post-doctoral fellowship in order to acquire some biochemistry so that he could research DNA properly. After some time with the biochemist Kalckar in Copenhagen, he moved on to England in the autumn of 1951, to the famous Cavendish Laboratory of Cambridge University. At this time, Sir Lawrence Bragg[2], one of the founders of crystallography and a specialist in X-ray diffraction studies was head of the laboratory. Here, Watson came into contact with a number of first-rate physicists and chemists (Max Perutz and John Kendrew among others) who were working on the elucidation of the three-dimensional structure of proteins by X-ray diffraction. Watson had recognized that a further structural analysis of DNA would have to be based on the correct interpretation of X-ray diffraction photographs of this compound.

1 This and the following quotes in the section "Nucleic Acids, the Key Substance of Living Matter" are taken from *The Double Helix* by James D. Watson, Atheneum, New York, 1968.
2 Bragg, Sir William Lawrence, 1890–1971, British physicist. Nobel Prize for Physics in 1915.

"I explained that I was ignorant of how X-rays diffract, but Max [Perutz] immediately put me at ease.
I was assured that no high-powered mathematics would be required..."
"I did not follow Max at all. I was even ignorant of Bragg's Law, the most basic of all crystallographic ideas."

Based on the experimental X-ray diffraction data concerning DNA which the physicist Maurice Wilkins and the crystallographer Rosalind Franklin were in the process of amassing in London at that time, and together with the physicist Francis Crick and the help of stimulating discussions with other chemists and crystallographers, Watson tried to solve the puzzle of the three-dimensional structure of DNA. At the same time, the world-famous structural chemist, Linus Pauling, was working on the same problem in the U.S.A. Pauling, who had discovered the α-helix structure of certain proteins shortly before, knew very well how important the discovery of the correct three-dimensional structure of DNA, this biologically most interesting molecule, would be.

"Helices were then at the center of the lab's interest, largely because of Pauling's α-helix.
Maurice [Wilkins] had told Francis [Crick], however, that the diameter of the DNA-molecule was thicker than would be the case if only one polynucleotide (a collection of nucleotides) chain were present. This made him think that the DNA molecule was a compound helix composed of several polynucleotide chains twisted about each other."
"Moreover, the black cross of reflections which dominated the picture [X-ray diffraction photograph] could arise only from a helical structure."

There developed a close race between the research groups in Great Britain and in the U.S.A. Watson and Crick used a method to determine the structure of DNA which Pauling himself had found very successful in the establishment of his α-helix, namely the use of atom and molecule models.

"The key to Pauling's success was his reliance on the simple laws of structural chemistry. The α-helix had not been found by only staring at X-ray pictures; the essential trick, instead, was to ask which atoms like to sit next to each other. In place of pencil and paper, the main working tools were a set of molecular models superficially resembling the toys of pre-school children."

Watson and Crick tinkered with their various models, always searching for a structure which would satisfy all the known chemical facts about DNA, would be stereochemically justifiable, would correspond to the data derived from the X-ray diffraction studies, and would suggest a reduplication mechanism for the DNA molecule.

"I had decided to build two chain models. Francis [Crick] would have to agree. Even though he was a physicist, he knew that important biological objects come in pairs."

Early in 1953, Pauling and Corey seemed to have come first in the race. They published their conclusions as to the three-dimensional structure of the DNA molecule. According to their paper, the structure involves three polynucleotide chains, wound around each other. The phosphate residues of the chains were supposed to be located near the vertical axis of this triple helix and the bases on the outside of the chains. Watson and Crick immediately noticed that this model from the acknowledged master of structural chemistry contained serious flaws. On April 2, 1953, the biologist Watson and the physicist Francis Crick[1] submitted to the journal *Nature* a paper (see Fig. 45) with a structural chemistry content, entitled: "A Structure for Deoxyribose Nucleic Acid" which contained the correct model of the three-dimensional structure of the DNA molecule.

According to it, deoxyribonucleic acid molecules consist of two polynucleotide chains which are arranged in a right-handed helix about an imaginary axis. The chains run anti-parallel. The phosphate residues

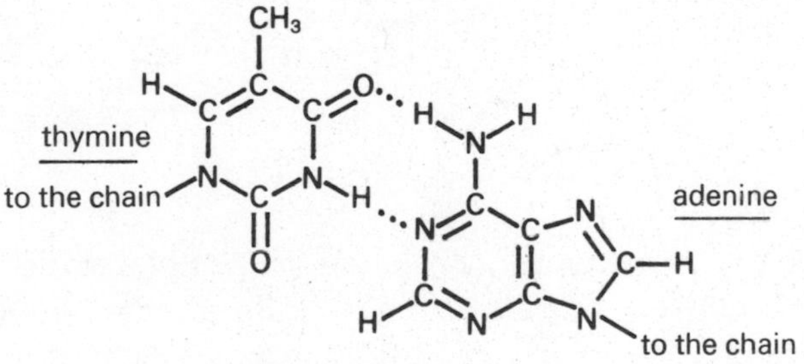

The base pairing of adenine and thymine in a DNA double helix molecule due to the formation of two hydrogen bonds. In the places marked "to the chain", the bases are connected to the deoxyribose component of their respective DNA strand.

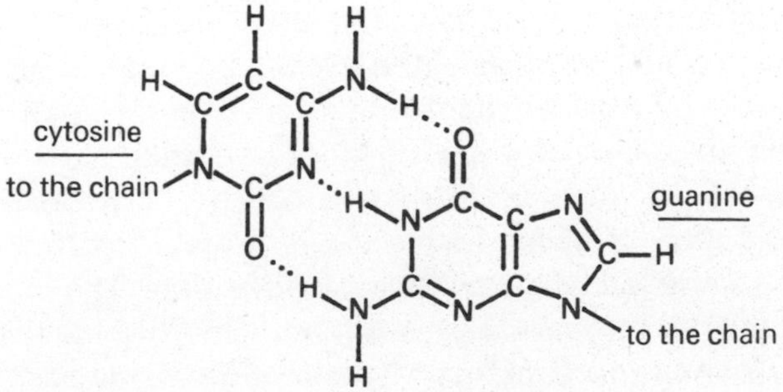

The base pairing of cytosine and guanine in a DNA double helix molecule due to the formation of three hydrogen bonds.

1 Crick, Francis, born 1916, British scientist. Nobel Prize for Physiology or Medicine in 1962, together with Maurice Wilkins and James D. Watson.

of the nucleotides point outwards, whereas the nitrogenous bases are found on the inside of the double helix. The bases perform an important function in the stabilization of the double helix. Certain bases of one helix pair with certain complementary bases of the other helix through the formation of hydrogen bonds in such a manner that a maximum stability for the molecule is achieved. In this bonding of bases it is always the bases adenine (A) and thymine (T), or guanine (G) and cytosine (C), which pair. The one base provides the pyrimidine component; the other always supplies a corresponding purine base for the pair and vice versa. The base pairs

148 *A section of a DNA double helix molecule showing base pairs T-A and C-G. Whereas the left DNA chain forms its phospho-di-ester bonds between the 3'-position and the 5'-position, the situation is exactly the reverse in the right DNA chain. The two DNA chains are antiparallel.*

are complementary. The two helices are thus not bonded on the principle of "like-to-like" (similarity) but according to the principle of complementarity. The plane of the base pairs is at right angles to the vertical axis of the double helix. The entire double helix arrangement looks like a spiral staircase, the steps of which are the base pairs. The reason why only bases A and T, or G and C, pair together is that members of such pairs are of the same shape, so that their pairing creates "steps" of a uniform size on the "staircase". Moreover, this combination of bases forms the strongest hydrogen bonds in comparison to the other possible combinations. The base pairs maintain a distance of 0.34 nm between them. One turn of the double helix measures 3.4 nm and contains 10 nucleotide residues.

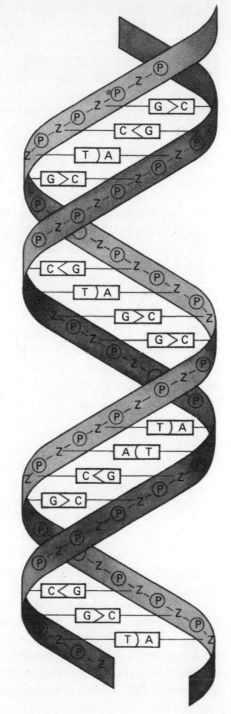

*The symbols have the
following meaning:*

P = phosphate residue
Z = sugar
  ≙ 2-deoxyribose
A = adenine
C = cytosine
G = guanine
T = thymine

(taken from spektrum,
No. 6, 4th Series, 1973.)

The Duplication of DNA

In their *Nature* article of 1953, Watson and Crick referred to the duplication problem in the following, mock-modest sentence: "It has not escaped our notice that the specific pairing [base pairing] we have postulated immediately suggests a possible copying mechanism for the genetic material." This mechanism, according to which the DNA in every living cell of every living organism is duplicated before cell-division takes place, proceeds in the following manner: First, a "parent" double helix unwinds, forming two polynucleotide chains. The base pair bonds are dissolved in this unwinding process. Thus the bases on each of the two polynucleotide strands are free to bond with complementary bases on the mononucleotides in the immediate cellular neighbourhood. Mononucleotides, which exist as energy-rich triphosphates in the cellular milieu, are bonded to the original strand according to the law of base pair complementarity to be polymerized into a new polynucleotide chain. The energy necessary for the polymerization is released when the triphosphates are transformed into the required monophosphates. In this manner, a new double helix, identical to the original double helix, is formed. The new "daughter double helix" thus consists of one strand from the original double helix and one, newly formed complementary strand. Since there were two original "parent" strands, two "daughter double helices" have been formed, and the DNA material in the cell has been duplicated. This mechanism, whereby the polynucleotide chains of the original DNA molecule are retained and can repeatedly participate in further new formations of complementary strands is known as semi-conservative replication. Both the despiralization of the DNA molecule and the polymerization of the newly bonded mononucleotides on these dissociated strands are cellular processes which require the action of enzymes.

Until now, we have spoken of deoxyribonucleic acids, or simply DNA, without giving any information as to the size or length or the degree of polymerization or the relative molecular weight of these polymers. It must be emphasized and remembered that there is no such thing as DNA in general; instead, there exist very many variants of natural DNA. Every type and species of organism possesses its own specific DNA or specific deoxyribonucleic acids. The various kinds of DNA are distinguished from each other by the difference in the number of nucleotide building blocks, in their quantitative relationship (how many of each kind of mononucleotide) and in the spe-

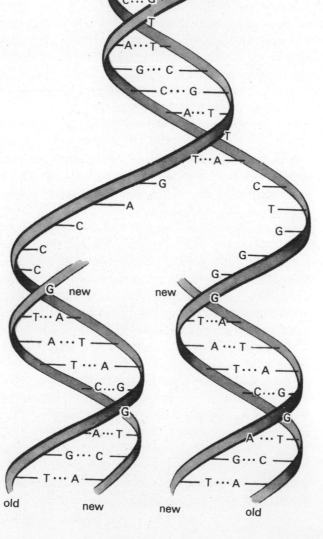

The replication of DNA.

After the despiralization of the parent DNA double helix, two single DNA strands are formed (parent-parent). Complementary deoxyribonucleotides attach themselves to the bases of these DNA strands and are there polymerized to form new polynucleotide strands. In this manner, two double helices come into being, each consisting of one parent strand and one new strand.

cific sequence of these building blocks. For instance, the *Escherichia coli* bacteriophage Lambda possesses a DNA double helix molecule with a molecular weight of 32 million. The length of a mammalian DNA double helix is estimated at 2 metres. The estimate is based on the assumption that each chain of the double helix represents a sequence of 5.5. billion nucleotides. Although it has been possible to extract DNA molecules of uniform lengths from bacteria and viruses, we have not as yet been notably successful in obtaining uniform natural DNA molecules from higher organisms. Attempts to do so have mostly produced only segments of the complete DNA. The reason for this failure is the fact that the DNA molecules from bacteria and viruses are relatively "small" and survive the process of extraction and preparation intact. The larger the DNA molecule, the more sensitive and fragile it is. In terms of the external shape of natural DNA, electron-microscopic photographs have shown that these molecules occur both in linear and in ring form. Furthermore, there have been reported observations of chain (catenane) structures (see Chapter II), i.e. of inter-linked DNA rings.

The Genetic Code

The replication of the genetic material is only one of the important functions performed by the DNA molecules. Another, equally important function is their role in the synthesis of proteins. The synthesis instructions for all the various proteins as well as exact details for every required polypeptide chain are stored in the DNA. It is the DNA that carries the information as to which amino-acids are to be bonded in what sequence into which polypeptide chain. A given segment of a strand of DNA contains the code, the instructions, for a particular polypeptide. This DNA segment or informational unit is what we know as a gene. A gene contains between 500 and 6,000 nucleotide pairs. However, polypeptides are not synthesized directly on the DNA strand. In eucaryotes, the DNA is, as a rule, concentrated in the nucleus whereas protein synthesis takes place in the cytoplasm[1] surrounding the nucleus. The "factories for the manufacture of protein" are the spherical to ellipsoid cell organelles called the ribosomes. These ribosomes are made up of ribonucleic acid (RNA) and proteins. In the 1960s, it was discovered that "messenger" molecules bring the information for the protein synthesis from the DNA to the ribosomes. These "messengers" are certain ribonucleic acids called

1 Cytoplasm: all cellular matter outside the cell nucleus in eucaryotic cells.

messenger-RNA or mRNA for short. The instructions encoded in the DNA are transcribed onto the mRNA. How is this transcription process effected? After the despiralization of the DNA strands, ribonucleotides attach themselves according to the law of base complementarity to the "free" bases of one DNA strand. They are then polymerized into an RNA chain. This complementary mRNA chain or strand then releases itself and travels to the ribosomes in the cytoplasm where it acts as the actual "blueprint" according to which the actual assembly of the polypeptide chains takes place. With the aid of other RNA types—the so-called transfer ribonucleic acids (tRNA)—the DNA information which has been transcribed onto the mRNA is decoded. This last-mentioned step of the process is known as translation. The rule that genetic information is irreversibly transcribed from the DNA to the RNA to be then translated into protein, that this process is intransitive, is the central dogma of molecular genetics.

"Virtually all the evidence then available made me believe that DNA was the template upon which RNA chains were made. In turn RNA chains were the likely candidates for the templates for protein synthesis.
On the wall above my desk I taped up a paper sheet saying DNA → RNA → Protein. The arrows did not signify chemical transformation but instead expressed the transfer of genetic information from the sequences of nucleotides in DNA molecules to the sequences of amino-acid proteins."

Recently, processes have been discovered in which the RNA functions as the matrix for the formation of DNA as well. However, the rule that information is never passed from the proteins to the DNA remains unshaken.

Let us now return to the question of the genetic code. A code is a rule or a set of rules which governs the translation of one sequence of signals or symbols into another, equivalent sequence of signals or symbols. For instance, the transcription of the letters of the Latin alphabet into the dots and dashes of the Morse code represents an encoding process. Whereas the Latin alphabet consists of 26 symbols (the letters, a–z), the Morse "alphabet" possesses only two symbols, the dot and the dash. By combining these two symbols into code units, it becomes possible to encode all 26 Latin alphabet symbols into Morse units; the a becomes .–, and l becomes .–.. in the Morse code. We can regard the relationship between nucleotides and amino-acids in an analogous way. Natural proteins are built of 20 different amino-acids, which means that the amino-acid alphabet requires 20 symbols.

We have already seen that natural DNA, as a rule, is constructed of four mononucleotides, namely dAMP, dGMP, dCMP, and dTMP. These four nucleotides are thus the available symbols for encoding the genetic information. Since the information is always transcribed from the bases of one strand of despiralized DNA, we can use the abbreviations for the four different bases as our code symbols, namely A, G, C, T.

If *one* base symbolized *one* amino-acid, then only four amino-acids could be encoded. If we started from the assumption that two bases would each serve as the coding unit for the amino-acids, then it would be possible to encode only $4^2 = 16$ amino-acids, a number which is four short of the required 20 amino-acids for natural proteins. We will not continue to play guessing games but state clearly the secret of the genetic code. The genetic code is based on code units (codons) consisting of three bases each. Each amino-acid is thus

154

The genetic code.

The individual dice contain the abbreviations of the 20 proteinogenic amino-acids and the designation "term" which is equivalent to a stop signal. A combination of three bases from the so-called "base-co-ordinates" gives us the codons for the amino-acids, e.g. AUG for methionine (taken from Naturwissenschaften, *7, 1978).*

symbolized by a combination of three bases; the code is a triplet code. Since there are four bases and the codons consist of only three bases each, the possible number of codons equals $4^3 = 64$, despite the fact that only 20 codons are required to encode the 20 amino-acids. In the genetic code we have the phenomenon that several nucleotide codons refer to one and the same amino-acid. We then speak of a degenerate code. 61 of these 64 codons symbolize amino-acids; the remaining three codons represent stop or punctuation symbols, also necessary in the construction of different polypeptide chains. These codons indicate the termination of the construction of a particular polypeptide chain. Further elucidation of the genetic code also answered the key question as to which nucleotide codons symbolize which amino-acids. The research that answered these questions was primarily carried out by the biochemists Nirenberg, Matthaei, Khorana, and Ochoa in the U.S.A. in the 1960s. Let us recapitulate. The information encoded in the DNA is transcribed onto the mRNA. RNA is chemically distinguished from DNA in that it has a ribose instead of a deoxyribose and the structurally nearly identical base uracil (it pairs with adenine, just like thymine) in place of the base thymine. Since the mRNA are the actual stencils, the matrices, for the construction of the polypeptide chains, the RNA bases A, G, C, U are used to symbolize the genetic code as it can be found transcribed on the mRNA. Thus, on mRNA, the amino-acid methionine is coded by AUG, phenylalanine by UUU or UUC, and arginine by any one of six different codons.

There is another important aspect to the genetic code which must be mentioned: it is valid for all living organisms. Monod, who developed the later experimentally validated hypothesis of the mRNA together with Jacob in 1961, put it in the following words:

"That which applies to *Escherichia coli*[1] also applies to the elephant."

The in vitro Synthesis and Modification of Genes

Since genes are chemical structures, biochemists have naturally been interested in synthesizing such "blueprints for life". Pioneer work in this area was done by the Indian Nobel Laureate, Har Gobind Khorana, and his co-workers in the U.S.A. By way of a truly inspired combination of methods from preparative organic chemistry plus enzymatic methods, Khorana in 1970 achieved the first complete laboratory synthesis of a gene of 77 nucleotide pairs. The gene which

155

1 *Escherichia coli:* an intestinal bacterium, the favoured research organism in molecular biology.

The total artificial synthesis of genes.

The drawing shows the four DNA duplexes (I–IV), which were used for the synthesis of a structural gene consisting of 126 nucleotide pairs. Moreover, the 26 oligonucleotide segments can be recognized by the bracketed numbers next to them. The individual mononucleotides which were used are here symbolized by the base abbreviations only.

156

(From H. G. Khorana et al., in: The Journal of Biological Chemistry, Vol. 251, No. 3.)

```
         26  25  24  23  22  21
        ┌───────────────(5)─────────
        T─ C─ C─ A─ A─ G─
                         T─ C─
        └───────────────────────────
```

```
    56  55  54  53  52  51  50  49  48  47  46  45  44  43
   ┌──────────────────(11)──────────────┐
    T─ C─ T─ G─ A─ G─ A─ T─ T─ T─ A─ G─ A─ C─
    │   │   │   │   │   │   │   │   │   │   │   │   │   │
 G─ A─ G─ C─ A─ G─ A─ C─ T─ C─ T─ A─ A─ A─ T─ C─ T─ G─
 └──────────(12)──────────┘           └──────────(10)──────────┘
```

```
    94  93  92  91  90  89  88  87  86  85  84  83  82  81  80  79  78  77
                       ┌──────────────────(18)──────────────┐
                        G─ G─ C─ A─ C─ C─ A─ C─ C─ C─ C─ A─ A─
                        │   │   │   │   │   │   │   │   │   │   │   │   │
 A─ T─ T─ A─ C─ C─ C─ G─ T─ G─ G─ T─ G─ G─ G─ G─ T─ T─
 └──────────(19)──────────┘           └──────────(17)──────────┘
```

```
    126 125 124 123 122 121 120 119 118 117 116 115 114 113 112 111 110
   ┌─────────(26)─────────┐           ┌──────────────────(24)──────────────
    C─ G─ A─ A─ G─ G─ G─ C─ T─ A─ T─ T─ C─ C─ C─ T─ C─
    G─ C─ T─ T─ C─ C─ C─ G─ A─ T─ A─ A─ G─ G─ G─ A─ G─
   └───────────────────────(25)───────────────────────┘
```

he synthesized is the one which codes for the alanine-transfer-RNA[1] of yeasts. Unfortunately, attempts to establish the biological effectiveness of the artificial gene were unsuccessful. When introduced into a living cell, it did not perform its function as a gene. A few years later, this last obstacle was also overcome although with a different artificially synthesized gene. This gene, also synthesized by Khorana and co-workers, consisted of 126 nucleotide pairs. This one was the structural gene for the precursor of the tyrosine-suppressor-tRNA in *Escherichia coli* bacteria. When introduced into a living cell, this artificial gene performed its function flawlessly.

What are the principles according to which these gene syntheses proceed? To begin with, deoxyribomononucleotides must be synthesized. These must then be joined, forming oligonucleotides of specific sequences by means of certain chemical methods. In the complete synthesis of the last-named structural gene, 26 oligonucleotide segments were synthesized chemically. The oligomeric segments consisted of 13 nucleotide building blocks at the most. These segments

---

1 Transfer-ribonucleic acids (tRNA) are ribonucleic acids which bring amino-acids to the ribosomes, to transfer them to the growing polypeptide chains. For every amino-acid necessary for protein synthesis there exist in the cell one or more specific tRNA to act as transport.

```
 20 19  18 17  16 15  14 13  12 11  10  9   8   7    6   5   4   3   2   1
                        (3)                              (1)
C - T - T - A - G - G - A - A - G - G - G - G - G - T - G - G - T - G - G - T   (5')        [I]
G - A - A - T - C - C - T - T - C - C - C - C - C - A - C - C - A - C - C - A   (3')
                (4)                                  (2)
```

```
 42 41  40 39  38 37  36 35  34 33  32   31 30  29 28  27  26 25  24 23
-(9)                              (7)
G - G - C - A - G - T - A - G - C - T - G - A - A - G - C - T   (5')        [II]
C - C - G - T - C - A - T - C - G - A - C - T - T - C - G - A - A - G - G - T   (3')
                        (8)                              (6)
```

```
 76 75  74 73  72 71  70 69  68 67  66 65  64 63  62 61  60 59  58 57
            (15)                              (13)
G - G - G - C - T - C - G - C - C - G - G - T - T - T - C - C - C - T - C - G   (5')        [III]
- C - C - C - G - A - G - C - G - G - C - C - A - A - A - G - G   (3')
        (16)                          (14)
```

```
109 108 107 106 105 104 103 102 101 100  99 98  97 96  95 94  93 92  91 90
                    (22)                              (20)
G - T - C - C - G - G - T - C - A - T - T - T - T - C - G - T - A - A  T  G   (5')        [IV]
- C - A - G - G - C - C - A - G - T - A - A - A - A - G - C   (3')
(23)                              (21)
```

were joined with the help of the enzyme polynucleotide ligase into four so-called DNA duplexes which contained corresponding complementary single-strand ends. The phosphorylation of the 5'-hydroxy groups of the overlapping single-strand ends prepared these four duplexes for a further joining into a complete gene, once more using the enzyme polynucleotide ligase for the joining process. In these steps, Khorana managed to produce about 1/1,000,000 (one millionth) of a gram of this structural gene. By far the largest amount of time expended in these gene syntheses is taken up by the purely chemical synthesis steps and not by the enzymatic procedures. A more economical structuring of the chemical synthesis steps will likely be an area of intensive research in preparative organic chemistry in the next few years.

Khorana's success with gene synthesis shows that it is quite possible, in principle, to produce truly artificial genes. This would mean not only artificial duplications of natural genes but also the synthesis of modified or completely original genes. The creation of original genetic structures and their introduction into certain organisms (gene transfer) is designed to alter the genetic constitution of these organisms. The idea is to develop organisms with certain planned new genetic traits.

Since the process for producing artificial genes by way of combined chemical and enzymatic methods is, at present, far too expensive, the currently more common experiments in this line of research proceed by way of extracting natural genes from organisms with various characteristics and then recombining these genes in an attempt to achieve a modified genetic material. When this "new" genetic material is then introduced into an organism, it acquires additional genetic material. In this manner it is possible, for instance, to alter the metabolism of micro-organisms in such a way as to increase or even initiate the production of desirable metabolites such as antibiotics, hormones, or other organic pharmaceuticals. This procedure opens new horizons in the development of bio-technological synthetic methods. Together with the methods designed to alter or create genetic material, come associated interests such as gene surgery and genetic manipulation, also known as genetic engineering. Genetic engineering is presently practised primarily on micro-organisms. For more complicated organisms there are as yet only very remote possibilities for undertaking gene manipulation. Manipulation of the human genome is not yet in sight.

The prospect of gene manipulation raises spectres as well as hopes. One danger, for instance, is the possibility that gene manipulation will produce new strains of bacteria and viruses which will be resistant to antibiotics. The proliferation of such organisms could give rise to new and dangerous diseases. In the distant future, there are also the dangers of a possible misuse of gene manipulation in the changing of the human genome. The positive hopes that scientists harbour in relation to gene manipulation concern the possible new strategies for combating hereditary diseases, viral infections and cancer.

39　Linus Pauling with the space-filling model of a helical structure.

159

Following page:
40/41   Space-filling models of D- and L-alanine, amino-acids with chirality.
Fig. 40: D-alanine; Fig. 41: L-alanine.

42   X-ray diffraction photograph of a myoglobin crystal. Myoglobin is the protein whose structural analysis by means of diffraction studies was carried out by J. C. Kendrew and his co-workers. In recent decades, crystallography has contributed decisively to the structural elucidation of complicated organic compounds, or, indeed, has made them possible in the first place. Examples of such structural analyses are that of penicillin, of vitamin B$_{12}$, of proteins and of nucleic acids.

43   Model of a haemoglobin molecule, which consists of two α-chains (white building blocks) and two β-chains (black building blocks). The flat grey plates represent the haeme groups.

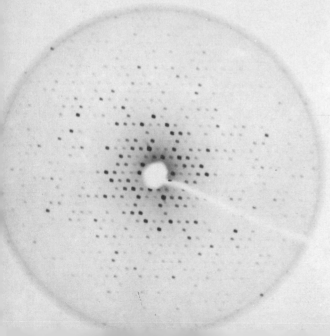

44 Molecule model of the tertiary structure of the enzyme bovine trypsin. This hydrolase enzyme is a protease, i.e. protein-degrading enzyme. It is constructed of 223 amino-acid residues which, altogether, contain 3,222 atoms.

162

equipment, and to Dr. G. E. R. Deacon and the captain and officers of R.R.S. *Discovery II* for their part in making the observations.

[1] Young, F. B., Gerrard, H., and Jevons, W., *Phil. Mag.*, **40**, 149 (1920).
[2] Longuet-Higgins, M. S., *Mon. Not. Roy. Astro. Soc., Geophys. Supp.*, **5**, 285 (1949).
[3] Von Arx, W. S., Woods Hole Papers in Phys. Oceanog. Meteor., **11** (3) (1950).
[4] Ekman, V. W., *Arkiv. Mat. Astron. Fysik. (Stockholm)*, **2** (11) (1905).

MOLECULAR STRUCTURE OF NUCLEIC ACIDS

A Structure for Deoxyribose Nucleic Acid

WE wish to suggest a structure for the salt of deoxyribose nucleic acid (D.N.A.). This structure has novel features which are of considerable biological interest.

A structure for nucleic acid has already been proposed by Pauling and Corey[1]. They kindly made their manuscript available to us in advance of publication. Their model consists of three intertwined chains, with the phosphates near the fibre axis, and the bases on the outside. In our opinion, this structure is unsatisfactory for two reasons: (1) We believe that the material which gives the X-ray diagrams is the salt, not the free acid. Without the acidic hydrogen atoms it is not clear what forces would hold the structure together, especially as the negatively charged phosphates near the axis will repel each other. (2) Some of the van der Waals distances appear to be too small.

Another three-chain structure has also been suggested by Fraser (in the press). In his model the phosphates are on the outside and the bases on the inside, linked together by hydrogen bonds. This structure as described is rather ill-defined, and for this reason we shall not comment on it.

We wish to put forward a radically different structure for the salt of deoxyribose nucleic acid. This structure has two helical chains each coiled round the same axis (see diagram). We have made the usual chemical assumptions, namely, that each chain consists of phosphate diester groups joining β-D-deoxyribofuranose residues with 3′,5′ linkages. The two chains (but not their bases) are related by a dyad perpendicular to the fibre axis. Both chains follow right-handed helices, but owing to the dyad the sequences of the atoms in the two chains run in opposite directions. Each chain loosely resembles Furberg's[2] model No. 1; that is, the bases are on the inside of the helix and the phosphates on the outside. The configuration of the sugar and the atoms near it is close to Furberg's 'standard configuration', the sugar being roughly perpendicular to the attached base. There

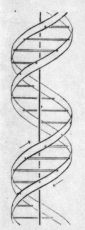

This figure is purely diagrammatic. The two ribbons symbolize the two phosphate—sugar chains, and the horizontal rods the pairs of bases holding the chains together. The vertical line marks the fibre axis

is a residue on each chain every 3·4 A. in the z-direction. We have assumed an angle of 36° between adjacent residues in the same chain, so that the structure repeats after 10 residues on each chain, that is, after 34 A. The distance of a phosphorus atom from the fibre axis is 10 A. As the phosphates are on the outside, cations have easy access to them.

The structure is an open one, and its water content is rather high. At lower water contents we would expect the bases to tilt so that the structure could become more compact.

The novel feature of the structure is the manner in which the two chains are held together by the purine and pyrimidine bases. The planes of the bases are perpendicular to the fibre axis. They are joined together in pairs, a single base from one chain being hydrogen-bonded to a single base from the other chain, so that the two lie side by side with identical z-co-ordinates. One of the pair must be a purine and the other a pyrimidine for bonding to occur. The hydrogen bonds are made as follows : purine position 1 to pyrimidine position 1 ; purine position 6 to pyrimidine position 6.

If it is assumed that the bases only occur in the structure in the most plausible tautomeric forms (that is, with the keto rather than the enol configurations) it is found that only specific pairs of bases can bond together. These pairs are : adenine (purine) with thymine (pyrimidine), and guanine (purine) with cytosine (pyrimidine).

In other words, if an adenine forms one member of a pair, on either chain, then on these assumptions the other member must be thymine ; similarly for guanine and cytosine. The sequence of bases on a single chain does not appear to be restricted in any way. However, if only specific pairs of bases can be formed, it follows that if the sequence of bases on one chain is given, then the sequence on the other chain is automatically determined.

It has been found experimentally[3,4] that the ratio of the amounts of adenine to thymine, and the ratio of guanine to cytosine, are always very close to unity for deoxyribose nucleic acid.

It is probably impossible to build this structure with a ribose sugar in place of the deoxyribose, as the extra oxygen atom would make too close a van der Waals contact.

The previously published X-ray data[5,6] on deoxyribose nucleic acid are insufficient for a rigorous test of our structure. So far as we can tell, it is roughly compatible with the experimental data, but it must be regarded as unproved until it has been checked against more exact results. Some of these are given in the following communications. We were not aware of the details of the results presented there when we devised our structure, which rests mainly though not entirely on published experimental data and stereochemical arguments.

It has not escaped our notice that the specific pairing we have postulated immediately suggests a possible copying mechanism for the genetic material.

Full details of the structure, including the conditions assumed in building it, together with a set of co-ordinates for the atoms, will be published elsewhere.

We are much indebted to Dr. Jerry Donohue for constant advice and criticism, especially on interatomic distances. We have also been stimulated by a knowledge of the general nature of the unpublished experimental results and ideas of Dr. M. H. F. Wilkins, Dr. R. E. Franklin and their co-workers at

King's College, London. One of us (J. D. W.) has been aided by a fellowship from the National Foundation for Infantile Paralysis.

J. D. WATSON
F. H. C. CRICK

Medical Research Council Unit for the
Study of the Molecular Structure of
Biological Systems,
Cavendish Laboratory, Cambridge.
April 2.

[1] Pauling, L., and Corey, R. B., *Nature*, **171**, 346 (1953) ; *Proc. U.S. Nat. Acad. Sci.*, **39**, 84 (1953).
[2] Furberg, S., *Acta Chem. Scand.*, **6**, 634 (1952).
[3] Chargaff, E., for references see Zamenhof, S., Brawerman, G., and Chargaff, E., *Biochim. et Biophys. Acta*, **9**, 402 (1952).
[4] Wyatt, G. R., *J. Gen. Physiol.*, **36**, 201 (1952).
[5] Astbury, W. T., Symp. Soc. Exp. Biol. 1, Nucleic Acid, 66 (Camb. Univ. Press, 1947).
[6] Wilkins, M. H. F., and Randall, J. T., *Biochim. et Biophys. Acta*, **10**, 192 (1953).

Molecular Structure of Deoxypentose Nucleic Acids

WHILE the biological properties of deoxypentose nucleic acid suggest a molecular structure containing great complexity, X-ray diffraction studies described here (cf. Astbury[1]) show the basic molecular configuration has great simplicity. The purpose of this communication is to describe, in a preliminary way, some of the experimental evidence for the polynucleotide chain configuration being helical, and existing in this form when in the natural state. A fuller account of the work will be published shortly.

The structure of deoxypentose nucleic acid is the same in all species (although the nitrogen base ratios alter considerably) in nucleoprotein, extracted or in cells, and in purified nucleate. The same linear group of polynucleotide chains may pack together parallel in different ways to give crystalline[1-3], semi-crystalline or paracrystalline material. In all cases the X-ray diffraction photograph consists of two regions, one determined largely by the regular spacing of nucleotides along the chain, and the other by the longer spacings of the chain configuration. The sequence of different nitrogen bases along the chain is not made visible.

Oriented paracrystalline deoxypentose nucleic acid ('structure B' in the following communication by Franklin and Gosling) gives a fibre diagram as shown in Fig. 1 (cf. ref. 4). Astbury suggested that the strong 3·4-A. reflexion corresponded to the internucleotide repeat along the fibre axis. The ∼ 34 A. layer lines, however, are not due to a repeat of a polynucleotide composition, but to the chain configuration repeat, which causes strong diffraction as the nucleotide chains have higher density than the interstitial water. The absence of reflexions on or near the meridian immediately suggests a helical structure with axis parallel to fibre length.

Diffraction by Helices

It may be shown[5] (also Stokes, unpublished) that the intensity distribution in the diffraction pattern of a series of points equally spaced along a helix is given by the squares of Bessel functions. A uniform continuous helix gives a series of layer lines of spacing corresponding to the helix pitch, the intensity distribution along the nth layer line being proportional to the square of J_n, the nth order Bessel function. A straight line may be drawn approximately through

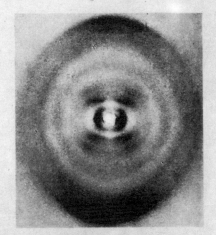

Fig. 1. Fibre diagram of deoxypentose nucleic acid from *B. coli.* Fibre axis vertical

the innermost maxima of each Bessel function and the origin. The angle this line makes with the equator is roughly equal to the angle between an element of the helix and the helix axis. If a unit repeats n times along the helix there will be a meridional reflexion (J_0^2) on the nth layer line. The helical configuration produces side-bands on this fundamental frequency, the effect[5] being to reproduce the intensity distribution about the origin around the new origin, on the nth layer line, corresponding to C in Fig. 2.

We will now briefly analyse in physical terms some of the effects of the shape and size of the repeat unit or nucleotide on the diffraction pattern. First, if the nucleotide consists of a unit having circular symmetry about an axis parallel to the helix axis, the whole diffraction pattern is modified by the form factor of the nucleotide. Second, if the nucleotide consists of a series of points on a radius at right-angles to the helix axis, the phases of radiation scattered by the helices of different diameter passing through each point are the same. Summation of the corresponding Bessel functions gives reinforcement for the inner-

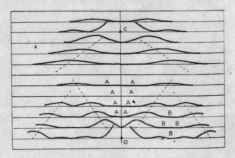

Fig. 2. Diffraction pattern of system of helices corresponding to structure of deoxypentose nucleic acid. The squares of Bessel functions are plotted about 0 on the equator and on the first, second, third and fifth layer lines for half of the nucleotide mass at 20 A. diameter and remainder distributed along a radius, the mass at a given radius being proportional to the radius. About C on the tenth layer line similar functions are plotted for an outer diameter of 12 A.

46   J. D. Watson (left) and
F. C. H. Crick at the Cavendish
Laboratory, Cambridge, in
front of their skeleton model
of DNA.

47   The ball and stick model
of DNA.

48   Har Gobind Khorana, born 1922.

49   Electron-microscopic photograph of the DNA ring of the simian virus.

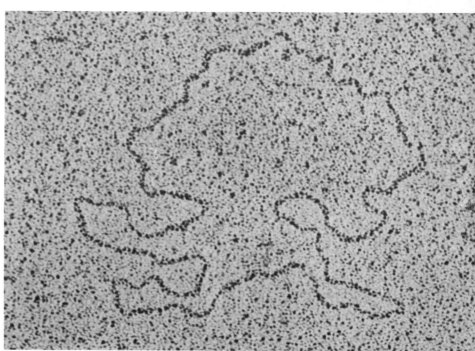

V.

The Uses of Chemistry

Chemistry is a science whose results, both practical and theoretical, are used in a variety of ways in order to fulfil the individual and social needs and desires of mankind. Our technological development would be at an entirely different stage if we had no fertilizers, pesticides, drugs, detergents, synthetics, fuels, paints, varnishes, or photographic materials. This wide range of industrially important chemical products clearly indicates the economic significance of chemistry and its considerable role in safeguarding our standards of living.

That the effects of chemistry can also be destructive rather than constructive is amply demonstrated by the use of explosives and toxic chemical compounds as weapons.

Various areas where applied chemistry makes a large contribution have been pointed out in the preceding chapters of this book. Let us only recall, at this point, things like diamond abrasives, graphite, the medically important vitamin B_{12}, the synthetic polymers (plastics, elastics, and synthetic fibres), and the possible uses of synthetic genes.

As a productive force, chemistry is continually involved in a variety of ways in the ongoing process of confronting and mastering the important problems which beset humanity. Among these problems we must count such global concerns as population control, the maintenance and re-establishment of human health, and the securing of an adequate food supply for a growing world population. Using these three categories in order to limit our area of consideration, we will try to illustrate some uses of chemistry in greater detail.

One chemical advance in organic chemistry in the 20th century, which must be considered as especially effective in social terms, is the development of oral contraceptives. If humanity should continue to increase at the present rate, then we estimate that the world population of about 4 billion (1977) will have doubled to 8 billion in the next 35 years.

The major increases in the world population take place in the underdeveloped countries of Asia, Africa, and South America, ever since the dramatic reduction of the mortality rate in these countries at the beginning of the 20th century. Experts estimate that it will not be possible to increase food production, especially in these underdeveloped countries, fast enough to keep pace with the hunger of the living and those yet to be born; we cannot guarantee them a life free from hunger. Consequently, there are worldwide efforts underway to reduce the population increase by decreasing the birth rate. Birth control involves a whole series of contraceptive measures of which the "pill" is only one.

At present (1977), of about 500 million women of childbearing age in the world, nearly 55 million are using the contraceptive "pill". This makes the pill the number one birth control method in the world, and it can be reasonably assumed that its use will increase in the years to come. The mechanism of action of the contraceptive pill involves the ingestion of so-called sex hormones. These hormones prevent the maturation and ovulation of oocytes in the female body by simulating the hormonal condition of pregnancy. A combination of sex hormones will, however, produce an apparent menstrual cycle despite the pseudo-pregnancy.

The development of hormonal contraception began about 1920. At that time, the physiologist Haberlandt managed to de-fertilize female rabbits by transplanting the ovaries of pregnant rabbits under

Oestrone (an Oestrogen)

Progesterone

their skin. The subsequent sterility of these rabbits could be traced back to the effects of the sex hormones.

Towards the end of the 1920s, chemists managed to isolate and analyze the structure of two female sex hormones, oestrone and progesterone. Experiments with animals showed that these sex hormones inhibit ovulation, i.e. prevent the ejection of a mature ovum from the ovaries of a female body, so that there is nothing to be fertilized. However, both these compounds have a serious disadvantage. Taken orally, they are rapidly metabolized. The resultant ovulation inhibition is only slight, a feature that made them unsuitable as a contraceptive per se. This fact led to research with the aim of synthesizing a "more stable" sex hormone. In 1938, the chemist Inhoffen and the biochemist Hohlweg, working in Berlin, reported the successful synthesis of an orally effective Oestrogen, the 17α-ethinyl-oestradiol, as well as that of an effective progesterone, the 17α-ethinyl-testosterone.

17α-ethinyl-oestradiol (an Oestrogen)

17α-ethinyl-testosterone

The general use of hormonal contraceptives began in 1959. In the course of testing and development, the progesterone which had been synthesized by Inhoffen and Hohlweg was modified slightly. The methyl group in position 10 was replaced by a hydrogen atom. This modified compound was actually dispensed under the name of norethynodrel as an oral contraceptive in the mid-1960s. Today, we have a number of different synthetic and practically effective progesterones which all derive structurally from the norethynodrel and which all have the ethinyl group in the 17α-position[1]. The Oestrogen synthesized by Inhoffen and Hohlweg, the 17α-ethinyl-oestradiol, or rather its 3-methyl ether, is also still a component of most oral contraceptives. The progesterone in the contraceptive inhibits ovulation whereas the Oestrogen induces a monthly bleeding despite the pseudopregnancy.

The synthesis of Oestrogens and progesterones is based on the compound oestrone. The Oestrogen 17α-ethinyl-oestradiol is synthe-

1 The 17α-position means that the ethinyl group in position 17 is in an α-position, i.e. it can be found behind the ring plane.

sized from oestrone in a one-step reaction process. Oestrone was first isolated from the urine of pregnant mares. However, as a raw material source for oestrone, mare urine is unsuitable since the amounts in the urine are minimal. Instead, a more advantageous source was found in the naturally occurring and plentiful sterines, like cholesterol and sistosterol. These steroles possess the basic structure of the sex hormones, so that only a partial synthesis is required in order to obtain oestrone. To begin with, 11 reaction steps were required to derive 17α-ethinyl-oestradiol from cholesterol. More recently, the use of micro-biological systems has drastically reduced the necessary number of reactions.

Another suitable, naturally occurring raw material source for the sex hormones is diosgenine, a compound which is found in the roots of a number of *Dioscorea* plant species. This plant family is especially widespread in Mexico, India, and the Peoples' Republic of China. Oestrone is also a suitable compound for synthesizing the progesterones.

Cholesterol

Diosgenine

Besides the extremely economical partial syntheses of sex-hormones from natural compounds which already contain the basic four-ring structure, the end of the 1960s also saw the development of a number of total syntheses which are also economically viable. It was necessary to solve various stereo-chemical problems in order to achieve these total syntheses. Chemists have thus provided the wherewithal, both with the partial and the total syntheses of the Oestrogens and the progesterones, to satisfy any increase in the future demand for hormonal contraceptives. However, it is true, and remains true, that birth control is not merely a matter of effective chemistry. It meets a great deal of resistance in precisely those countries which need it most. Some of the many reasons for this resistance are poverty, poor health care systems, and religious taboos.

170

Chemistry is of primary importance both for the maintenance and the restoration of human health. The diagnosis, prevention and/or fight against many diseases would be impossible without the knowledge, the methods, and the products supplied by chemistry. In close interdisciplinary association with biology and medicine, chemistry, in the form of pharmaceutical chemistry, contributes decisively to both drug research and production. Chemistry is directly involved on various levels in the development and manufacture of all kinds of pharmaceuticals. Such is particularly the case in:

- The extraction of therapeutically effective natural substances such as the alkaloids, vitamins, hormones, antibiotics, etc. from plant and animal sources;
- the structural analysis of these compounds and the development of partial or total syntheses in order to make manufacture feasible (see Chapter II, The Total Synthesis of Vitamin B_{12});
- the manufacture of derivatives from these natural compounds (see the section on penicillin below);
- the synthesis of original or modified active agents;
- the development of analytical test methods for such active agents and their metabolites;
- the development of the technological requirements for the manufacture of drugs.

Synthetic chemistry makes an exceedingly important contribution to both pharmaceutical research and production. Most of the chemical compounds tested for their medical efficacy are not natural substances but synthetic compounds which originated in the chemical laboratories. An annual testing of several hundred thousand compounds for their biological effectiveness presupposes an immense programme of chemical synthesis. These syntheses involve both the manufacture of known and original compounds. If a compound is synthesized that shows biological activity, that class or related classes of compounds will then become the subject of further research.

A recent example of such spin-off effect is the development of silicon pharmaceuticals, a series of organosilicon substances. In 1963, the Soviet scientist M. G. Voronkov discovered that arylsilitranes are exceedingly toxic. This led to the synthesis of a large number of related compounds. Some of them turned out to possess physiological activity and showed a variety of therapeutic effects. It seems likely that they

171

will find their place in the pharmacopaeia. Among them, we find migugen, an ethoxysilatrane, which has a projected use in cancer chemotherapy.

Migugen (ethoxysilatrane)

172

A great deal of synthetic work is expended in the alteration of the chemical structure of known drugs. Sometimes, very slight modifications of the constitution or the configuration of a compound can provide a considerably improved product in terms of its effectiveness, both qualitative and quantitative. Such modifications also sometimes increase the therapeutic range of the original compound.

The chemical or microbial synthesis methods developed in the laboratories for the synthesis of these compounds usually form the basis for the development of technology for their mass production. Synthetic chemistry is thus also concerned in the developing of ever more efficient and economical synthetic methods for drugs which are already in production and in adapting the existing processes for changing raw materials.

The end of the 19th century and the whole of the 20th have been remarkable for great progress in the development of a large number of effective drugs. Examples of such drugs are the narcotics, the analgesics, the psychiatric drugs, the cardiovascular drugs, the blood substitutes, the vitamins, the hormones, the antiseptics, the sulphonamides, the antibiotics, the cytostatics, the virostatics, the antimycotics, and others. Effective anaesthetics, narcotics, and antiseptics are indispensable in modern surgery. Mental illnesses can be treated with the psychiatric drugs which have been developed in the most recent decades. Without analgesics, people would be obliged to endure considerably more pain. A large variety of deficiency ailments and conditions caused by lack of vitamins or hormones can now be cured or at least controlled thanks to the availability of these compounds. Cytostatics together with surgery and radiation techniques are used in the fight against cancer. The use of antibiotics, antimycotics, virostatics and sulphonamides has helped defeat a pleth-

Trapanal—a modern narcotic which is injected.

Veronal (barbiturate)—a sleeping pill.

Novocaine—the most frequently used local anaesthetic.

Aspirin—a mild analgesic.

173

Dolantin—a strong analgesic.

Vitamin C—an effective agent against scurvy and a prophylactic against the common cold.
Isonicotinic acid hydrazide—an effective drug against tuberculosis.

Meprobamate—a psychotropic drug.

Salvarsan—the first highly effective chemical treatment for syphilis.

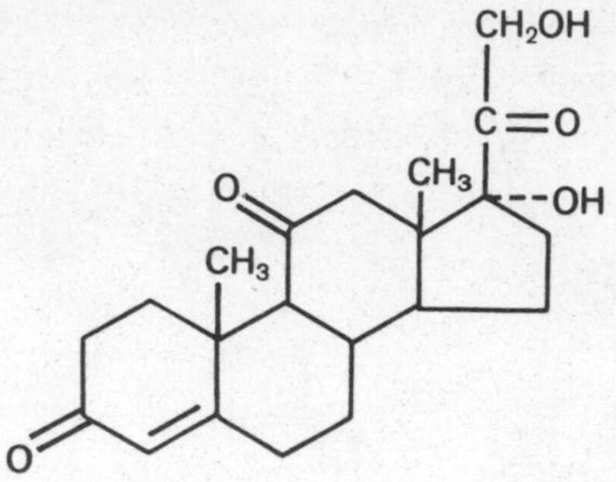

Cortisone—used in rheumatism therapy and for other things.

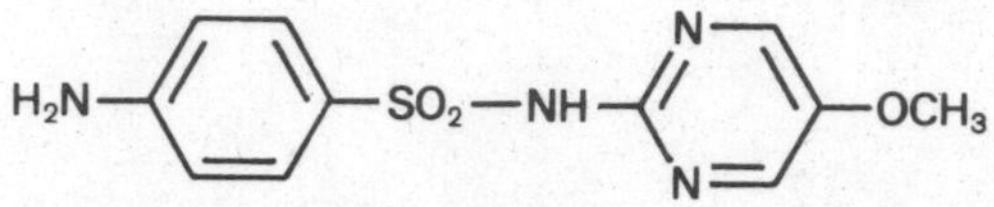

Sulphamethoxydiazine—a modern sulphonamide for the treatment of bacterial infections.

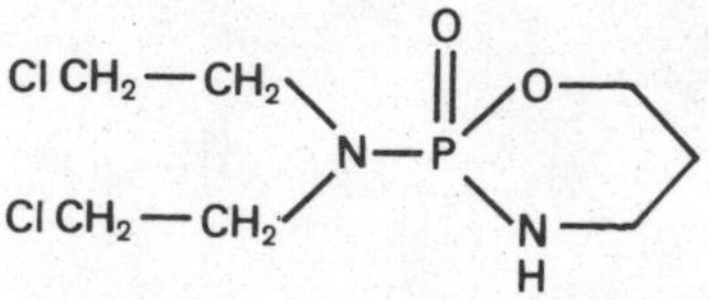

Cyclophosphamide—a drug used in cancer chemotherapy.

ora of infectious diseases caused by micro-organisms. Examples such as tuberculosis, dysentery, typhoid, venereal diseases, and others immediately spring to mind.

Among the modern drugs, the antibiotics form an especially important group. Their isolation, purification, structural analysis, partial and total synthesis and, finally, their usually biotechnological mass production require a constant high-quality input from chemical research and chemical process technology. Antibiotics are those compounds produced by micro-organisms which possess the ability to kill other micro-organisms or to prevent or inhibit their proliferation. The best known group of antibiotics are the penicillins.

In 1929, Alexander Fleming[1] published a scientific paper in which he described the antibiotic effects of the mould *Penicillium notatum* on a series of pathogenic bacteria such as staphylococci and streptococci. Fleming came to the conclusion that the mould produces a chemical compound which has this antibiotic effect. He named this effective compound penicillin, deriving the name from that of the mould involved, namely *Penicillium notatum* WESTLING. He did not, however, succeed in isolating the responsible chemical compound from the nutrient medium for the mould at this time. One of the reasons for this failure is the high sensitivity of the penicillins to chemical

1 Fleming, Sir Alexander, 1881–1955, British bacteriologist. Nobel Prize for Physiology or Medicine in 1945, together with Ernst B. Chain and Howard W. Florey.

agents and to heat. Another is the low concentration of the compound in the culture medium. It took another ten years for the scientists Florey[1] (a pathologist) and Chain[2] (a chemist) and others working together with Fleming developed the necessary techniques for the successful isolation and purification of penicillin, an absolute prerequisite for the clinical testing of the compound. Fleming described the reasons for the long delay in the isolation of penicillin in the following reminiscence:

> "It may have been a great misfortune that ten years passed between the time that I discovered penicillin and the time at which it could be purified in sufficient quantity and concentration so that its therapeutic properties could be at least partially exploited. As far as I am concerned, the reason for the delay is simple: I am a bacteriologist, not a chemist, and I was working in a bacteriological laboratory with no experienced resident chemist ... We made no progress because our amateurish efforts ... to concentrate it [penicillin] ... had little success, and since penicillin is a labile substance, it was often already inactive when we tested it."[3]

World War II forced the pace of research on penicillin both in Great Britain and in the U.S.A. It also caused a high priority to be given to the development of mass production methods for penicillin since this compound was regarded as *the* miracle drug in the fight against wound infections. The structural analysis of the penicillin which Fleming had discovered in Great Britain, the so-called penicillin F, yielded the following structural formula:

Penicillin F = pentenyl-penicillin

pentenyl

Similar analyses carried out in the U.S.A. on the there developed penicillin yielded another formula. A slightly different penicillium mould had been used, producing penicillin G, or benzyl-penicillin (see p. 176).
Synthesis chemists have developed total syntheses for the natural penicillins but these cannot compete economically with the production of the antibiotics by their micro-organisms. A much more important development was the synthesis of many semi-synthetic penicillins. The triumphant progress of the antibiotics against the infec-

1 Florey, Sir Howard Walter, 1898–1968, British pathologist. Nobel Prize for Physiology or Medicine in 1945 together with Alexander Fleming and Ernst Chain.
2 Chain, Sir Ernst Boris, 1906–1979, British biochemist (born in Germany). Nobel Prize for Physiology or Medicine in 1945 together with Alexander Fleming and Howard Florey.
3 *Via Gloriosa*, Vienna, 1956, pp. 310–311.

benzyl

tious diseases encountered one serious obstacle: the use of anti-biotics has shown that the micro-organisms which they are to destroy will develop a resistance or tolerance to the particular antibiotic that is being employed against them. This problem is solved over and over again (since it arises over and over again) by modifying known antibiotics chemically within certain limits, and by continuous research for new antibiotics to which the pathogenic micro-organisms have not as yet developed resistance. With penicillin, the production of these semi-synthetic variants proceeds as follows. Large amounts of penicillin are produced with the help of huge cultures of penicillium mould. The side chains of the harvested penicillin are then split off with the help of a specific enzyme. In this manner, large amounts of the "basic compound" of the penicillins, the 6-amino-penicillanic acid, which in itself is too inactive to be useful as a drug, are produced.

6-amino-penicillanic acid

This compound is the basic raw material for the manufacture of all the semi-synthetic penicillins. Using "normal" chemical synthesis methods, the basic structural unit is provided with side chains on the $^+NH_3^-$ group and the side chain possibilities are extremely varied. In 1970, about 5,000 of such semi-synthetic penicillins were known, but only very few of them are ever dispensed in general medical practice since only very few display the required properties in sufficient quality and quantity. One of the ones that turned out to be very effective is ampicillin.

In 1943, S. A. Waksman discovered streptomycin, another important antibiotic. This is the antibiotic of choice for treating tuberculosis. Normally, the antibiotics are a product of a fermentation process, i.e. micro-organisms are used to synthesize them. However, there is one antibiotic, chloramphenicol, which is mass produced in total chemical synthesis. Its relatively simple structure makes a purely chemical synthesis very economical.

$$O_2N-C_6H_3-CH-CH-CH_2OH$$

with NH—COCHCl_2, OH

Chloramphenicol is especially effective against typhoid and against the entire staphylococci family of micro-organisms which are resistant to other antibiotics such as the penicillins, the streptomycins, and the tetracyclins.

Recent decades have seen marvellous advances in our continuing battle against disease with the help of new chemical compounds. However, this does not mean that there are no more important problems for pharmaceutical research to resolve. On the contrary. In the developing countries, for instance, there remain many unsolved health problems in the treatment of many different parasite-induced diseases. Examples of such diseases are malaria, bilharzia, filaria, sleeping-sickness, leprosy, and others. Bilharzia and filaria, both worm-caused diseases, each affect between 500 and 600 million people. The battle against malaria has been partially successful. The method for trying to control it has been indirect, namely the annihilation of parasite carriers, the mosquitoes, with DDT. But it has been a very limited success. We still have hundreds of millions of cases of malaria every year. Also, the drugs and chemical agents which have been developed against these diseases are frequently in very limited use because the economic situation of the countries where the disease is rampant does not permit them to pay for an adequate programme. Also, available drugs are often only a compromise solution requiring much more research. Apart from the above-named priorities in pharmaceutical research other important research fields at present are: cardiovascular drugs, drugs for the treatment of arthritic conditions, mental diseases, and effective chemotherapy against malignant tumors.

As far as cancer is concerned, there have, indeed, been considerable advances in chemotherapy in the last thirty years. However, *the* effective drug against cancer has not been discovered. Such a drug would have to be a compound which would only attack those cells which are proliferating in an uncontrolled manner (the cancer cells) and leave normal cells strictly alone. The search for powerful cytostatics is being intensively pursued in research laboratories all over the world.

The discovery and introduction of a new drug is an exceedingly expensive process. Today, the situation is such that the developmental time span for a new effective drug is between 5 and 10 years and the price tag is between 20 and 30 million dollars. This is primarily due to the fact that very little is known about the basic relationship between chemical structure and biological activity. Basic pharmaceutical research is primarily concerned with elucidating these basic relationships so that it will be possible in the future to reduce the present high input of time and money into a largely random search.

Chemistry and the World Food Supply

Estimates show that, in 1976, 1 billion people in the world were inadequately nourished. The populations of the developing countries are particularly vulnerable to malnutrition and sheer hunger caused by a fluctuating food supply. And the situation remains unchanged despite the fact that a global implementation of the most modern methods of agriculture could feed all these people without any trouble. The differences in the development of agricultural production become clear if we consider the following figures. In the mid-1970s the average world production of wheat per hectar was 17 dt. However, in some areas of Europe agricultural units produced already 100 dt per hectar in the same period.

An essential prerequisite for the achievement of high harvests is a fertilization programme. Every harvest removes nutrients from the soil which fertilization must replenish if future harvests are not to diminish. The use of natural fertilizer is not enough if soil cultivation is very intensive.

It was the German chemist Justus von Liebig (1803–1873) who laid the scientific foundations for fertilization programmes in the last century and who became one of the founding fathers of modern plant nutrition and fertilization theory. It was he who proved that the then current ideas concerning plant nutrition were erroneous and helped to propagate the correct theory, namely that plants require mineral compounds, i.e. inorganic substances, for proper growth.

Besides water and carbon dioxide, the nutritional requirements of plants consist of the elements nitrogen, phosphorus, potassium, magnesium, and calcium as their macro-nutrients. As micro-nutrients, they require borium, copper, manganese, molybdenum, and others. These elements are absorbed by the plant in the form of certain inorganic compounds. A most important process in plant nutrition is the incorporation of nitrogen (N), which is present in huge amounts in the atmosphere in its gaseous form as N_2. Unfortunately, most plants cannot assimilate nitrogen from the atmosphere. The usual means is by way of ammonium ions (NH_4^+) or nitrate ions (NO_3^-) which can be derived from the common chemical compound ammonia (NH_3).

Of the cultivated plants, only the legumes, which include beans, peas, and lentils, can supply their nitrogen needs directly from the atmosphere. The nitrogen is utilized by way of ammonia. The actual "synthesizer" of the ammonia in this process is a bacterium which lives in symbiosis with legumes in the so-called "nodes" on the roots of these plants. Ammonia is the key substance from which plants derive their nitrogen.

Justus von Liebig, 1803–1873. **179**

It was one of the truly phenomenal successes of chemical research when a method for the industrial, economically acceptable mass synthesis of ammonia from hydrogen and nitrogen was developed, an absolute prerequisite for the mass production of nitrogen fertilizers. In this synthesis, hydrogen and nitrogen are brought to react with one another in the presence of catalysts at a temperature between 475° and 600 °C and under high pressure, about 20 MPa:

$$N_2 + 3H_2 \rightleftharpoons 2NH_3$$

The physico-chemical groundwork for this synthesis was carried out by the chemist Fritz Haber (1868–1934) in the years 1905–1910. The actual methodology, the technological prerequisites for the industrial mass production of ammonia, was largely the work of the chemist and chemical engineer Carl Bosch (1874–1940) in the years 1908–1913. In 1913, BASF (Badische Anilin und Sodafabrik) in Oppau near Ludwigshafen began with a production of 30 tons of ammonia per day. During World War I, the synthetic ammonia which was produced in Germany was little used for the manufacture of fertilizer; instead, it was turned into nitric acid (HNO_3) and used for the manufacture of explosives for the war effort.

The technology of ammonia synthesis has been further refined in the last few decades. Examples of improvements can be found in the use of turbo-compressors in order to increase the pressure to 35 MPa, and in the perfecting of the ammonia catalyst. The space-time exploitation rate in the high pressure reactors is between 10 to 15 times as high as that in less sophisticated systems.

180 *Fritz Haber, 1868–1934.*

The quantitative development in the global production of ammonia has been tremendous. Today, apart from sulphuric acid, ammonia is the most important inorganic compound to be synthesized and 90 % of all nitrogen production is based on this synthesis. In 1980, an estimated 100 million tons of nitrogen will be produced in the form of ammonia. From this ammonia, the following main nitrogenous fertilizers are derived: ammonium sulphate $(NH_4)_2SO_4$, urea $(NH_2)_2CO$, calcium-ammonium saltpetre—a mixture of calcium carbonate $CaCO_3$ and ammonium nitrate NH_4NO_3, and sodium nitrate $NaNO_3$. Liquid ammonia is also used as a fertilizer nowadays, since ammonia can be liquefied at room temperature at a pressure of only 0.8–0.9 MPa. Liquid ammonia has the highest nitrogen content of all nitrogen fertilizers, namely 82.4 % by weight, and is rapidly absorbed by the soil. Chemical research and technology has also contributed substantially to the development and production of other kinds of fertilizers, such as phosphate and potassium fertilizers.

Apart from inorganic or mineral fertilizers, essential efficient scientifically-based agriculture is the use of chemical plant protection. At present about one third of the potential global harvest is consumed or destroyed by deleterious organisms. If we were to stop using chemical control agents entirely, the amount destroyed would double. The agricultural plant protection chemicals divide into three main classes: the herbicides, the insecticides, and the fungicides.

The use of herbicides for the control and destruction of weeds and undesirable grasses in large plantations of sugar beets or grains results in the saving of human labour for cultivation and also helps to increase the harvest by eliminating competition for the minerals supplied by the fertilizers. Modern herbicides are organic compounds like the growth herbicide 2,4-D, which is used to control unwanted grasses in grain plantings. 2,4-D is the abbreviation for 2,4-dichloro-phenoxy-acetic acid.

This compound, which generally finds usage in its ester form, causes a disturbance in the hormone balance of the weeds; the equilibrium between nutrient absorption and nutrient usage is disturbed and the weeds literally "grow themselves to death".

In the 1960s, the production of herbicides in the U.S.A. achieved high values since the US Army was using these compounds for criminal military purposes, namely the chemical defoliation of the Vietnamese jungle. This ecological warfare was carried out by spraying a mixture of the butyl esters of 2,4-dichloro-phenoxy-acetic acid, and of the 2,4,5-trichloro-phenoxy-acetic acid, the "agent orange".

2,4 D = 2,4-dichloro-phenoxy-acetic acid

Insecticides are chemical compounds capable of destroying mature insects or their developmental stages which eat or otherwise damage crops. These species include potato bugs, fruit worms, aphids, tree borers, etc. The biological demands which we make on an insecticide are extremely complex, not to say contradictory. The insecticide is to destroy the undesirable insects, but it is not to harm the beneficial ones; it is also to be harmless to the crop plant and, by implication, to humans and domestic animals. The difficulty of fulfilling all these requirements is clearly illustrated by the history of DDT, the globally renowned and infamous insecticide. This compound, known since 1874, is 1,1,1-trichloro-2,2-bis(4-chlorophenyl)ethane whose insect destroying properties were discovered in 1939 by the Swiss chemist P. H. Müller (Nobel Laureate for Medicin or Physiology in 1948). This contact insecticide was rapidly deployed on a global scale. It was used in agriculture, forest management, and against household pests of all kinds, like flies, bed bugs, lice, moths, etc.

DDT dichloro-diphenyl-trichloro-ethane

The use of DDT came into question when the development of insect strains resistant to this chemical was discovered. These new strains could tolerate the insecticide, even in large doses. The phenomenon of resistance, not confined to DDT, can only be combatted by the continuous development of newer and newer types of insecticides.

The notoriety of DDT as opposed to its fame began when highly sophisticated measurements (by means of mass spectrometry) showed that DDT was ubiquitous all over the world. Traces of it could even be found in the ice of Antarctica. It also showed up, and could be shown to accumulate, in the fatty tissues of man and animals. The reason for this ubiquity of DDT can be found in the extraordinary stability of this compound. It resists decomposition and is not affected by weather or micro-organisms. Its decomposition into innocuous compounds is so slow that it generally is incorporated unchanged into the human and animal organism by way of the food chain. When this fact was established and became well known, a heated dispute concerning the possible toxic effects of DDT began. Towards the end of the 1960s, this resulted in the prohibition of DDT in some countries and in severe restrictions of its usage in others.

Toxic residues, even if only in trace amounts, are a general problem that accompanies the use of any chemical in plant protection. The problem continues to exist and is equally valid for the insecticides at present in widespread use, namely, the organophosphorus compounds. These new insecticides are largely constituted of phosphoric and thiophosphoric acid esters; the compounds are highly toxic to insects but, unfortunately, also for warm-blooded animals. However, compared with DDT, they decompose quite rapidly in the soil and in plants. Among the organophosphorus insecticides, the systemic insecticides have become the most important. These compounds are absorbed and distribute themselves throughout the plant. Insects which try to feed on the so-treated plant ingest the poison and are severely damaged or destroyed.

The development of the phosphoric acid esters as insecticides is primarily due to the German chemist G. Schrader, who in 1944 synthe-

182

Carbicron®— a systemic insecticide

sized the first effective compounds in this class. His name is also associated with the synthesis of such highly toxic organo-phosphorus compounds such as tabun, sarin, and soman, toxins which were developed in Germany before and during World War II as part of an arsenal for a possible chemical warfare. The peril of these chemicals lies in the fact that they can be inhaled in the gaseous form without any kind of warning irritation. An amount of less than 1 mg is fatal for a human being. At the end of the war, the German Army was in possession of a stockpile of 400,000 kg of sarin.

Organophosphorus compounds for chemical warfare.

tabun   sarin   soman

In our battle against undesirable insects, we are now in the process of developing new, more specific methods. One of these is the production of chemical sterilization agents which aim at the sterilization or reduction of fertility of specific harmful insects. Further research aims at the isolation and synthesis of sexual attractants of insects. Such compounds would allure a particular kind of insect and thus make possible its selective annihilation.

The fungicides, the third class of chemicals used in plant protection, is a category of compounds for combatting diseases in plants which can be traced back to fungus or mould infections. Examples of such plant diseases are cacao cancer, rice rust, grape mildew, and potato blight. In this field, the most extensive research currently under way is concerned with the development of systemic fungicides.

In general terms, it must be said that a ban on or complete abstention from the use of chemicals for pest and weed control is not really feasible if we would also deal with the problem of an inadequate food supply for a growing world population. The problem of toxic residues in the soil and in the food chain is not as acute with many of the newer compounds used for plant protection because they are not as stable as DDT. There is, in fact, no reason for panic-stricken prohibitions of the use of such chemicals. It is, however, of first importance to keep the ecological stress produced by toxic chemicals in the environment to an absolute minimum. Ways of doing this would be:

— the development of ecologically harmless pesticides, herbicides and fungicides;
— the careful, scientifically based application of these substances; i.e. the right chemical, in the right amount, at the right time, and in the right place;
— the continuous monitoring of the accumulation of residues both in the environment and in the tissues of animals and humans;
— the development of biological ways and means of insect and disease control in plants; e.g. the use of beneficial insects to control the pests.

Not only agriculture but also animal husbandry owes a great deal to the chemical industry. Nearly all the feed additives such as antibiotics, synthetic essential amino-acids, vitamins, urea, certain minerals and compounds containing trace elements are provided by the chemical industry. Urea is especially important for the natural synthesis of animal protein.

The largest single problem in the nutrition of many populations is that of supplying them with sufficient amounts of high-quality proteins. The usual source of supply is the production of meat, i.e. of animal protein. In order to encourage the growth of this protein, the usual feed for meat animals consists of plant proteins and/or animal protein in the form of fish meal. However, resources of these raw materials are limited, both in their production and in their availability. For instance, the manufacture of the high quality protein fodder, soy-bran concentrate, is limited to a few countries and any large-scale use by another country would involve a considerable degree of economic dependence on the supplying country.

In the search for protein fodder substitutes, urea has been found to be a possible source of nitrogen. In the stomachs of various ruminant animals there exist many types of bacteria and protozoa which can utilize simple nitrogen compounds such as urea, and, if carbohydrates are present as well, can use the components of these substances to build their own body-specific proteins, which in turn become the raw material for the ruminant's synthesis of its own protein.

During the 1930s, German farmers made their first attempts at cattle feeding with a fodder mixture consisting of dessicated sugar beet chips, molasses[1] (from sugar beets), and urea. Today urea is also mixed with silages or even straw. Since the late 1940s, the use of urea as a fodder additive has spread to other countries and expanded

184

1 Molasses and dessicated beet chips are waste products in the manufacture of sugar (saccharose) from sugar beets. They are used primarily as a carbohydrate source. Molasses contains an average of 50 % saccharose.

immensely. The use of urea as a nitrogenous fertilizer, as well as its use as the base for the manufacture of certain plasts and as an economical fodder additive has led to the construction of large facilities for the mass production of urea in recent decades. The technical production of urea is not, however, based on Wöhler's synthesis, but on a synthesis developed by A. Basarov[1]. Following a suggestion of his doctoral advisor, H. Kolbe (Leipzig, p. 37), Basarov developed a synthesis of urea from the simple inorganic compounds carbon dioxide and ammonia in 1870.

The reaction pattern for this synthesis is as follows:

$$2NH_3 + CO_2 \rightleftharpoons NH_4OCONH_2 \rightleftharpoons H_2N-\underset{\underset{O}{\|}}{C}-NH_2 + H_2O$$

ammonium
carbamate   urea

Conditions required for the mass production of urea with this reaction are a pressure of 15 MPa and a temperature of 150 °C to 180 °C. This is only one example of the possible reaction parameters. Under these conditions the reaction produces an intermediate ammonium carbamate, which becomes urea in a further reaction in which water is split off. Today, industry is already into an integrated ammonia and urea production in which the raw materials oxygen, nitrogen, methane and water are used, with ammonia and urea as the end products. 85 % to 90 % of the present-day urea production is used as fertilizer; 8 % to 11 % is used for various industrial purposes; 2 % to 4 % as a fodder additive. The estimated top per cent of 4 % fodder usage seems a very low percentage of the total. However, in absolute amounts, 4 % of an annual production of 30,000 kt urea (in 1976) comes to 1,200,000 t, a considerable amount.

A further possibility for closing the "protein gap" involves work with the microbial production of proteins. The principle behind this is the production of protein through the use of monocellular microorganisms such as yeasts or bacteria which multiply rapidly under controlled advantageous conditions. The incredible multiplication rate of yeasts and bacteria compared with other organisms can be seen in table 5:

1 Basarov, Alexander von, of Russian origin, born 1845 in Wiesbaden, died 1907 in St. Petersburg.

Table 5

Organisms	Weight doubling time	Organisms	Weight doubling time
Yeasts and bacteria	$1/2$ to 2 hours	chicks	2 to 4 weeks
algae	1 to 48 hours	pigs	4 to 6 weeks
grasses	1 to 2 weeks	cattle	1 to 2 months

(According to K. Weissermel, *Umschau 76* (1976), No 16, p. 513.)

If a nutrient solution containing 100 g dextrose and nitrogenous nutrient salts is innoculated with Torula yeast, up to 210 g yeast will develop. Of this biomass, 25 % is dry weight, half of it pure protein. Because of its origin, this protein is known as "single-cell protein". Single-cell protein derived from bacteria is usually more valuable in biological terms than the single-cell protein derived from yeasts. However, industrial mass production of yeast proteins was quantitatively developed to a higher degree so that, in 1977, yeast protein production was 800,000 tons annually.

For the microbial synthesis of protein, the following conditions must be met: carbon-rich raw materials, a stream of air, and an aqueous solution of nutrient salts with a high concentration of ions such as NH_4^+, K^+, Mg^{2+}, PO_4^{3-}, SO_4^{2-}, and Cl^-. The metabolic processes which produce protein, among other substances, from these raw materials is known as *fermentation*. The multiplication of micro-organisms is carried out both in laboratories, in small amounts, and on an industrial scale in large reactors.

The microbiological synthesis of proteins is a controlled chemical process. Chemical methodology and technology is used in the large-scale technical control of the process and for the necessary preparation techniques of the resultant product (by separation and purification). The combination of microbiology and chemical technology gave rise to a new science called biotechnology. This new field is at present in a state of rapid expansion, involving not only the manufacture of single-cell proteins, but also the production of antibiotics, enzymes, steroids, vitamins, amino-acids, citric acid, food dyes, insecticides, certain biopolymers, and even of the basic chemical compounds for industrial reactions. Biotechnology requires the pooled efforts of a large team of experts working together in an interdisciplinary manner. It requires help from microbiology, chemistry, biochemistry, genetics, biology, molecular biology, engineering, medical research, etc.

The economic manufacture of biomass through microbiological synthesis requires a cheap and plentiful source of carbon. For decades now, spent sulphite liquors[1] and molasses have been used as such a carbon source.

The use of crude oil or crude oil products as the carbon source for the microbiological synthesis of protein is not a new idea. In 1913 Söhngen, from the Institute for Microbiology of the Technical University of Delft, published a paper entitled *Petrol, Petroleum, Paraffin Oil, and Paraffin as Carbon and Energy Sources for Microbes.* In his research bacteria and other microbes had actually been grown on these substrates but the experiment did not lead to any industrial developments. It was not until Just, Schnabel, and Ullmann at the Institute for Fermentation Research in Berlin in 1949–50 discovered that certain yeasts can utilize crude oil as a carbon source that various countries began to place a high priority on the research and development of this new use for crude oil or crude oil fractions. The results of this research became industrial reality during the late 1960s and early 1970s with the construction of the first mass production plants. The processes employed in these plants use the alkanes contained in the above named substrates as the carbon source in protein biosynthesis.

At present, intensive research is under way in order to develop bacterial protein synthesis on the basis of natural gas. The usable alkane in natural gas is methane (CH_4). Another promising potential substrate for a future bacterial protein synthesis is methanol (CH_3OH), an inexpensive synthetic product. There is also a great deal of research being done on the possible utilization of cellulose waste products as a carbon source.

The advantages of the industrial mass production of biomass proteins are the following:

— the proportion of protein in the biomass is very high;
— it is possible to use cheap raw materials or even waste products as substrates;
— biomass production is independent of climate and location;
— production is continual and not seasonal;
— very little land is required for the plant;
— reliance on the import of foreign soya beans and grain can be reduced or eliminated in many countries.

Biomass is primarily used as a fodder additive in animal husbandry, a process which transforms it into palatable food for humans. A direct

1 Sulphite liquors: waste products in the manufacture of cellulose from wood pulp via the sulphite process. They usually contain a certain percentage of carbohydrates and acetic acid which can be utilized by the yeast *Torula utilis,* for instance, as a carbon source.

use of biomass in human nutrition has not as yet been achieved and there remain a number of unsolved problems. The main problem concerns digestibility or harmlessness of the other compounds besides proteins which are also contained in the biomass. Other problems are the taste and texture of the biomass. At present, biomass comes in the form of a yellowish powder that has no similarity to the meat or other sources of protein to which mankind is accustomed. Efforts continue to improve the texture of microbial protein by using the methods of the chemical fibre industry in order to spin the protein into fibrous products with a meat-like consistency. With the addition of artificial flavours, manufacturers hope to produce an acceptable meat substitute.

188

Future Tasks for Chemistry

The maintenance and development of the material basis of human civilization continues to pose ever new problems and make ever new demands of the science of chemistry. We must try to respond to the demands and to master the problems by using the considerable fund of knowledge which we have inherited and by developing new strategies through intensive and motivated research.

There are two sets of problems which have surfaced as being particularly complicated and threatening for all of mankind, and of which we have become acutely conscious in the last few years. These problems concern our diminishing resources of raw materials and our endangered environment.

The manufacture of the various industrial products requires raw materials. Since the demand for the consumables is growing more and more raw materials are needed. All the chemical elements and compounds required by the various world economies as raw materials exist in adequate amounts but in extremely varying availability and concentration. The extraction of these resources is dependent upon the expense and effort involved and on whether they can be reached with the available technology. Thus, the availability of resources is directly related to the state of our scientific knowledge and technology.

For instance, sea water is a huge resource reservoir containing nearly all the natural elements. However, only a few raw materials (such as sodium chloride and magnesium salts) are at present being extracted from sea water. We know that all the oceans of this earth together contain more than 6 million tons of gold but we lack the technological means for extracting this treasure in an economically justifiable way. In the future mankind will be forced to satisfy some of its resource requirements from sea water and chemistry must supply economically rational methods and technologies for meeting these needs.

Another possibility for supplying the growing need for resources is an advance into the deeper reaches of the earth's crust. However, the further one must go in order to find natural gas, crude oil, or the various metal ores, the higher the cost.

From the point of view of chemical industry, the main source of raw materials is nowadays provided by crude oil, natural gas, and coal. More than 90% of the industry's organic products are based on crude oil. However, the major part (circa 80 to 85%) of the world's crude oil production is used to create energy and not as a raw material. This use of oil as a fuel is totally irresponsible. The burning of natural gas and oil means that the component elements are thereupon irretrievably lost to human use. Mendeleyev's observation that to burn oil is to burn money is as valid today as it was then. Present estimates predict a severe shortage of natural gas and oil within 30 to 40 years. What can be done? To begin with, it is essential that the use of these two carbon sources for energy production be drastically restricted. This would necessitate a more extensive use of coal for energy production (coal reserves are 10 times greater than those of oil and gas) and the development of nuclear and solar energy. A more thorough exploitation of the compounds contained within crude oil would again demand the full ingenuity and co-operation of chemistry.

Although our coal reserves are massive, it is, nonetheless, ultimately a shameful waste to simply burn them. Considering the rapid depletion of our resources of oil and natural gas, it is clear that our present oil- and gas-based chemical industry will have to at least start converting to coal by the turn of the century at the latest. It will be possible to draw on our past knowledge and technologies of coal exploitation to some degree but we will require further development and new strategies in order to achieve a more efficient material exploitation of coal than we have achieved up to now. There are numerous problems in this area which chemical technology in particular is being called upon to solve.

As is shown by the example of coal, gas, and oil, some raw material problems can be solved through the substitution of one resource for another. Further avenues for the amelioration of or even solution to raw material problems are: a reduction in the specific material input for certain products; the complex and thorough exploitation of each raw material; the use of waste products as secondary raw materials; the development of recycling systems. The contribution of chemistry to the solution of these raw material problems will be a constant challenge for the imagination and ingenuity of chemists everywhere.

The complex problem of the protection and the maintenance of an adequate environment also needs the input of chemistry, and that to a considerable degree. The chemical industry itself is one of the sources of environmental pollution. No one needs to be reminded that rivers have been polluted by industrial waste and that the air we breathe has been polluted by exhaust gases from large chemical plants. It must be said that this type of pollution could today be eliminated or at least severely reduced. The technology and processes for the control of industrial pollution have been available for some time. However, these technical solutions to the pollution problem are expensive. For this, and a variety of other reasons, pollution control measures are not being implemented with the desirable consistency and dispatch.

A further source of environmental pollution are the chemical products or their residues. Among these are the herbicides and insecticides, the detergents in sewage, the lead compounds in fuel, and even the fluorinated hydrocarbons which are used in aerosol cans. To develop chemical products without these pollutant side effects or, at least, with less damaging side effects, is one of the foremost priorities of chemical research.

As our brief survey of the problems involving raw material supply and environmental pollution has shown, there is no shortage of socially significant problems or directions for chemistry, for today and for the future.

50   Chemistry and physical chemistry contribute via medical diagnostics to the maintenance and restoration of human health. The photograph shows the direct monitoring of the potassium ion concentration in the blood of a patient during a heart valve operation. The apparatus is a through-flow system which determines potentiometrically ion concentrations with ion-selective electrodes.

Right: The potassium through-flow measuring system.
Left: End of the operation table.
Middle: Anesthesia apparatus and material.
Front: Parts of the heart-lung machine.

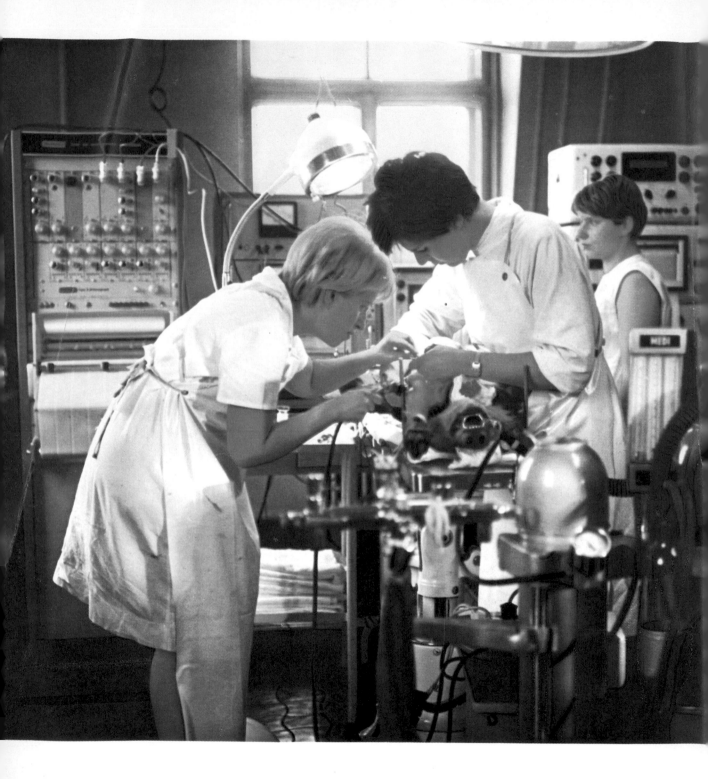

51 Testing of circulatory effectiveness of a newly developed chemical compound on a dog in a laboratory.

52 Penicillin producing mould culture (Penicillium notatum). The penicillin which is produced is given off into the surrounding culture medium and there inhibits the growth of bacteria, which can be seen in the organism-free zone around the mould culture.

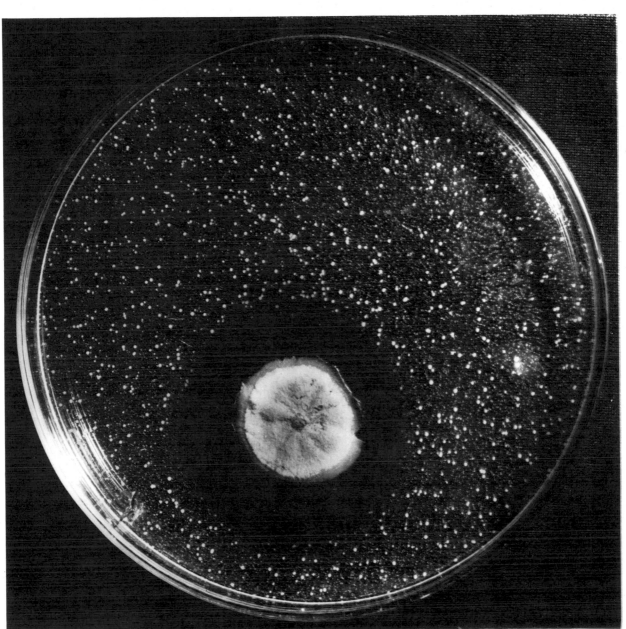

53 Small-sized fermentation equipment for producing antibiotics.

54 The tetracyclins are an important group of antibiotics. The photograph shows (left to right) the Soviet scientists Korobko, Kolossov and Guryevich with a molecular model of tetracyclin. These scientists achieved the first total synthesis of this basic structure of the tetracyclins.

55 Liebig's laboratory in Giessen.

56 *Photograph of the experimental apparatus used by F. Haber for the synthesis of ammonia from nitrogen and hydrogen. The three column-shaped parts of the apparatus are (from the right) a drying cylinder, a contact cylinder (reactor) and a liquefier respectively.*

57 A large industrial plant for the mass production of ammonia.

58 Interior of a fertilizer silo. The fertilizer in question is ®Nitrophoska, a complete fertilizer for both farming and garden use.

59   The destruction of the mangrove jungle in Vietnam, the result of the ecological warfare with herbicides which was carried out by the U.S. Army. The photograph was taken in 1970, several years after the actual "herbicide attack".

60   The effectiveness of the herbicide Pyramin in turnip cultivation.
Left: untreated; right: treated.

61   Large industrial plant for the mass production of urea.

62 The cooling of urea
granules in a layer of turbul-
ence.
The crystals which are produced in the synthesis of urea
have the disadvantage that
they stick together when
stored. For this reason the
crystals are melted, the liquid
melt forced through very
fine nozzles and turned into
a spray and the melt droplets then solidified through
cooling with a current of air
moving in the opposite direction. In this manner, granules
of 1 to 2.5 nm in diameter
are formed.

63 Droplets of oil on water,
colonized by yeast cells. The
oil in question is a crude oil
destillate.

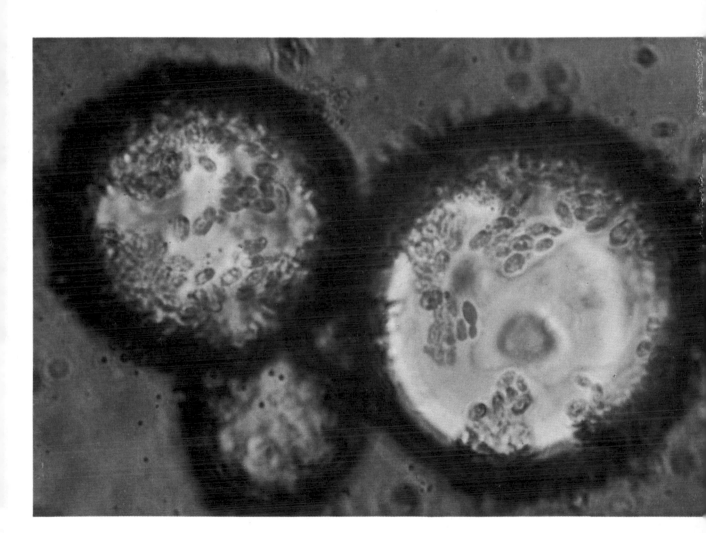

204

Appendix

Index of Names and Subjects

213

Sources of illustrations

ADN-Zentralbild, Berlin 25, 34, 51, 54, 61, 65

Archiv für Kunst und Geschichte, Berlin (West) 55

BASF Aktiengesellschaft, Presse und Information, Ludwigshafen 26, 27, 37, 38, 56, 58, 60, 62, 66

Bavaria-Verlag, Munich (Dr. Lorenz) 20, 21

Bild der Wissenschaft/Kage, Stuttgart 18

British Crown Copyright, Science Museum, London 47

The California Institute of Technology, Pasadena, California, U.S.A. 39

Camera Press Ltd., London 46

Chemie in unserer Zeit 2/67, Verlag Chemie GmbH, Weinheim 24

VEB Chemiefaserwerk "Friedrich Engels", Premnitz 33, 36

VEB Chemiekombinat Bitterfeld 28

Degussa, Frankfurt/Main 32

Deutsche Fotothek, Dresden 1

Deutsches Museum München 3

Deutsches Röntgen-Museum, Remscheid 19

DPA-Deutsche Presse Agentur GmbH, Frankfurt/Main 48

Düngemittelkombinat VEB Stickstoffwerk Piesteritz 57

Eidgenössische Technische Hochschule, Laboratorium für Organische Chemie, Zurich 50

Institut für Polymerenchemie der Akademie der Wissenschaften der DDR, Berlin 35

Christian Krebs, Berlin 4, 5, 6, 7, 9, 10, 11, 12, 13, 14, 15, 16, 30, 31, 40, 41

Laboratory of Molecular Biology, Cambridge, Great Britain 42, 43

Joachim Lachmann, Zentralinstitut für Physikalische Chemie der Akademie der Wissenschaften der DDR, Berlin 22, 23

Nature 4/53, London 45

VEB Petrolchemisches Kombinat, Schwedt 63, 64

VEB Reifenkombinat Fürstenwalde 29

Rijksmuseum voor de Geschiedenis van de Natuurwaetenschappen en van de Geneeskunde, Leiden 8

spektrum 9/76, Akademie der Wissenschaften der DDR, Berlin 53

Umschau in Wissenschaft und Technik, Frankfurt/Main 44

Arthur H. Westing, Hampshire College, Amherst/Massachusetts 59

Wissenschaft und Fortschritt, Akademie-Verlag, Berlin 49

Zentralinstitut für Mikrobiologie und experimentelle Therapie der Akademie der Wissenschaften der DDR, Jena 52

Zentralinstitut für Physikalische Chemie der Akademie der Wissenschaften der DDR, Berlin 17

Zentralstelle für Rationalisierungsmittel der Lehreraus- und -weiterbildung der Pädagogischen Hochschule "Dr. Theodor Neubauer", Erfurt 2

Illustrations in the text part:

Basilius-Presse AG, Basel p. 91

Archiv für Kunst und Geschichte, Berlin (West), pp. 15, 27, 36, 89, 115, 179

Deutsche Fotothek, Dresden pp. 13, 17, 18, 21, 22, 75

spektrum 4/78, Akademie der Wissenschaften der DDR, Berlin p. 119